Bioethics Across the Globe

Akira Akabayashi

Bioethics Across the Globe

Rebirthing Bioethics

 Springer Open

Akira Akabayashi
Department of Biomedical Ethics
University of Tokyo, Faculty of Medicine
Tokyo
Japan

This book is an open access publication.
ISBN 978-981-15-3571-0   ISBN 978-981-15-3572-7   (eBook)
https://doi.org/10.1007/978-981-15-3572-7

This Springer imprint is published by the registered company Springer Nature Singapore Pte Ltd.
The registered company address is: 152 Beach Road, #21-01/04 Gateway East, Singapore 189721, Singapore

To Aru, the only one who will inherit my genes.

Acknowledgments

I offer my sincere thanks to the many who offered scholarly insight at various points in preparing this book. Specifically, I thank Drs. Deborah Zion, Victoria University, and Eisuke Nakazawa, the University of Tokyo. I also thank Mr. Katsumi Mori, Ms. Fumiko Kan, and Ms. Asuka Naramoto of the University of Tokyo for their technical assistance. In addition, I really appreciate the continuous support of my family while I was writing this book: Aru, Mahoko, Noriko, and Kan. The publication of this book was partly supported by grants from Pfizer Health Research Foundation and Mitsubishi Foundation. Copyright: Images on pages, 116, 117, Getty Images; Images on pages 28, 95, Chinatsu Hagino; Images on pages, 102, 103, Asahi Shimbun Photo Archive.

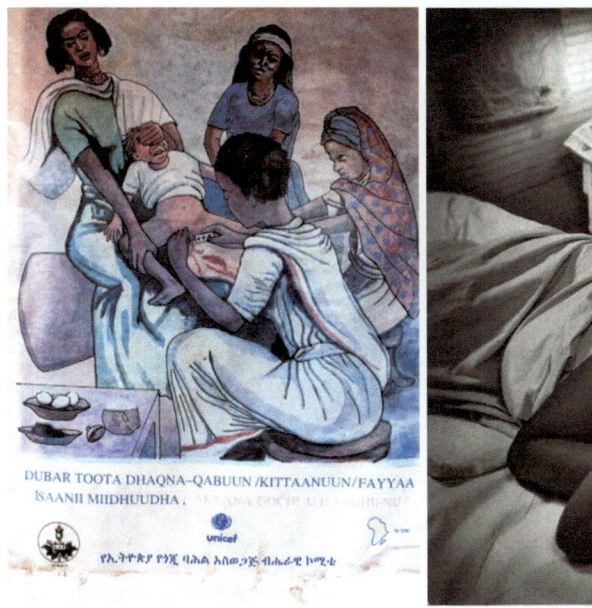

DUBAR TOOTA DHAQNA–QABUUN /KITTAANUUN/FAYYAA
ISAANII MIIDHUUDHA .

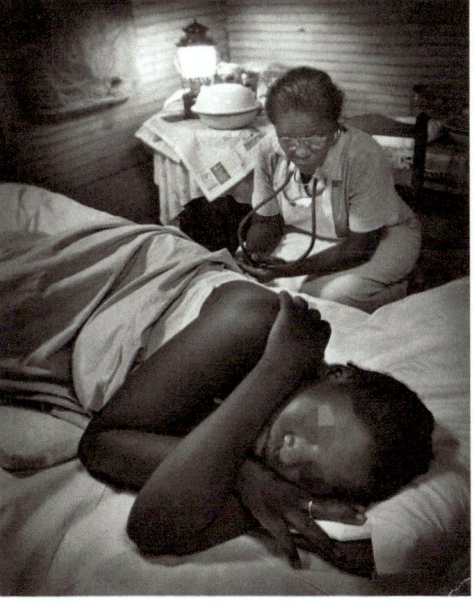

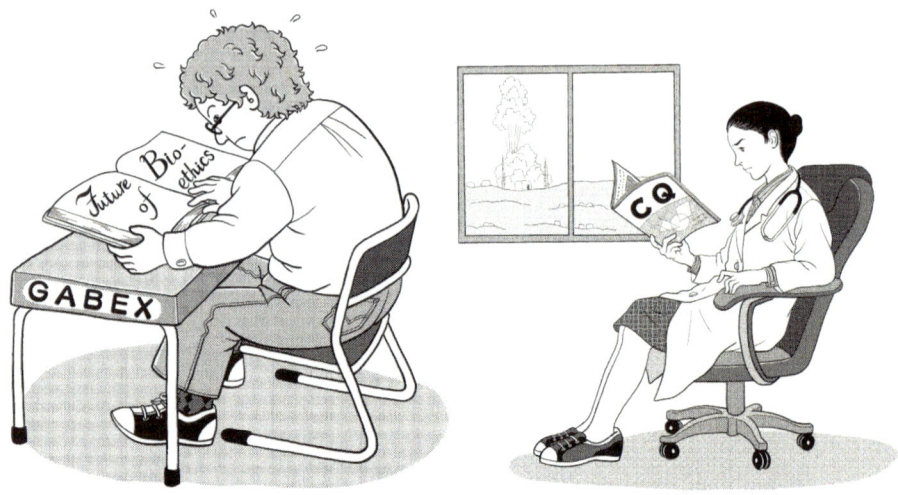

Introduction

Global bioethics has a short history. Van Rensselaer Potter was the first person who coined the term bioethics back in 1970, using it to describe the global "science of survival" [1, 2]. The term global bioethics has a shorter history and is now used widely in academic journals, such as the *Journal of Global Bioethics,* as well as in many books, including the *Encyclopedia of Global Bioethics* [3] and the *Handbook of Global Bioethics* [4]. The term is also used by nonprofit organizations such as the *Global Bioethics Initiative*. Notably, most of the conceptual frameworks used within this field such as "human rights," "human dignity," and "distributive justice" are all part of a Western ethical paradigm. In their foundational paper, Potter and Potter [5] note that the goal of global bioethics is "acceptable survival." They state that "(a) acceptable survival is a long-term concept with a moral constraint: worldwide human dignity, human rights, human health, and a moral constraint on human fertility." Human dignity and human rights are two key terms used in Western thinking, but it is unclear whether these concepts will be relevant in the future.

In the preface to *The Future of Bioethics: International Dialogues* [6], I wrote the following:

> Most studies in bioethics advocating East-West dialogue have either attempted cross-cultural comparison or proposed Eastern philosophical paradigms as a counter to Western ideas. The tacit premise of previous writing on East-West dialogue is therefore a strain of relativism. From the Eastern perspective, Western views are treated as a cultural construct that should be referenced as models, but are not appropriate to be utilized in their existing form. To Westerners, Eastern interpretation represents ways of thinking that should be recognized but can never truly be understood in their complexity within Western cultures. For this reason, Asians place Western conceptions of bioethics on the critical chopping block, and approach them as something to be overcome. In contrast, although Westerners occasionally comment on current conditions in Asian countries, they rarely fully engage with bioethical discussions led by Asian researchers, and neither express agreement nor fully critique such views. In a globalized world, simply maintaining a respectful distance from other cultures is no longer sufficient.

Furthermore, as stated by the former president of International Association of Bioethics (Campbell AV) [7], "(b)ioethics is not merely a Western philosophical pastime, but is a discipline that unites East and West, North and South in a common quest for solutions to the countless moral dilemmas of modern medicine and evolving biological sciences." Thus "global bioethics recognizes the importance of the local while thinking globally" [8]. It is therefore imperative to begin to listen more intently to the thoughts and perspectives about the values of those in other countries, regions, and cultures. The time has come for us to engage in an active discussion of our different cultural perspectives, in which attentive listening, rather than simply hearing, is the main objective.

One scholar of Japanese cultural studies wrote the following [9]:

> Ever since Japan formed its own civilization…, it has consistently been plagued by the fate of comparing and scrutinizing its own civilization with those of other continents... Beginning with literature, and ranging through policies, philosophy and religion, Japan has borrowed from other countries ….. in such a way that its own customs and cultures are often changed, and its own frameworks are processed so that they can coexist or fuse with others.… After the Meiji era, the target against which Japan compared itself simply changed from China to the West; the exact same process was adapted and has been applied continuously to this day. (Translated by the author)

Bioethics was born in the West—more accurately, the United States—during the second half of the twentieth century. The present text shares with the reader Japan's encounter with this Western concept of bioethics and its cohort of ethical issues, and how it has found its position as it decides where it stands, what it should accept, and what it has rejected (either openly or silently). Beginning with a brief history of bioethics in Japan, I introduce a variety of topics. I use my own writings and weave specific experiences from Japan into the narrative. This book may become a valuable source of information for those specializing in Asian or Japanese research, but it is *in no way* meant to be an introductory text about Japan itself.

In the last chapter, I present my own views on current global bioethics discussion. I go so far as to propose discarding the overarching term of Global Bioethics. Instead, I propose a more challenging term to accommodate future discussion. True Global Bioethics has yet to be born, and the longer we continue to limit ourselves to the use of concepts and methodologies from the West alone, the longer Global Bioethics will remain in stasis.

As many scholars have proposed, a dialogue that encourages both local and global thinking is needed. I believe that such a dialogue must be enabled by mutual understanding or, at the very least, a healthy attitude and sincere effort toward obtaining it. This book is intended to serve as a tool to promote this. Once the book has been read, the reader's understanding of Japan will be deeper than when he or she began.

Lastly, but most importantly, this book is open access. Many of the articles cited in this text are also open access. Thus, readers from LMICs (low to middle income countries) as well as students and laypersons can read them for free.

I urge all who read this book to write their own story for their own country to add to this dialogue. In this regard, this is not an introductory book on Japan, but rather the beginning of the series comprising many other narratives to come which will serve as tools that will facilitate the international and mutual understanding we require to initiate genuine dialogue.

- As noted above, I cite many of my own papers in the text. Many of them are open access, which means that they are available for free on the Internet. I have tried to highlight the essence of those papers in this text. I recommend that readers read the book before reading the open-access articles in depth, as this will facilitate a much better understanding of the overall narrative. To read open-access papers, from Google, go into "PubMed" and search for "Akabayashi." The free articles will be marked in orange. Clicking on the title will present the Abstract, and the full text can be read by clicking on "free article." PDF files can be downloaded if necessary.

References

1. Potter VR. Global bioethics: building on the Leopold legacy. Michigan University Press: Michigan; 1988.
2. Potter VR. Getting to the year 3000: can global bioethics overcome evolutions fatal flaw? Perspect Biol Med. 1990; 34:89–97.
3. Encyclopedia of Global Bioethics ten Have H ed. Switzerland: Springer International Publishing; 2016.
4. Handbook of Global Bioethics ten Have H, Gordijn B eds. Dordrecht: Springer Science + Business Media; 2014.
5. Potter VR, Potter L. Global bioethics: converting sustainable development to global survival. Med Glob Surv. 1995; 2:185–90.
6. Akabayashi A. Preface. In Akabayashi A, editor. The future of bioethics: international dialogues, Oxford: Oxford University Press; 2014. pp. v–vi.
7. Campbell AV. Presidential address: global bioethics – dream or nightmare? Bioethics 1999; 13:183–190.
8. Widdows H, Dickenson D, Hellstein S. Global bioethics. New Rev Bioeth. 2003; 1:101–116.
9. Okubo T. Genealogy of theory of Japanese culture. Chuo-Kohronsha, Tokyo, 2003, pp. i–ii. (in Japanese).

Contents

About the Author

Akira Akabayashi, MD, PhD, is a Professor in the Department of Biomedical Ethics at the University of Tokyo, Faculty of Medicine, Tokyo, Japan, and an Adjunct Professor of the Division of Medical Ethics at New York University School of Medicine, New York, NY, USA.
His research interests span cross-cultural bioethics, global bioethics, medical/clinical ethics such as informed consent, organ transplantation, end-of-life issues, medical humanities, including professionalism, public health ethics, research ethics, and bioethics policy making.

As an academic researcher, he has published more than 200 original articles and more than 20 books/textbooks or chapters in English, in addition to many Japanese publications. These include the English textbook *Biomedical Ethics in Asia: A Casebook for Multicultural Learners*, McGraw-Hill Education, 2010. As an educator, Professor Akabayashi has taught bioethics/medical ethics for over 25 years. Professor Akabayashi has served as the chair of the Research Ethics Committee at the Kyoto University Faculty of Medicine and the University of Tokyo Faculty of Medicine for more than 15 years. He is also the Founding Director of Office for Human Research Studies where he conducts research ethics consultation at the University of Tokyo. Professor Akabayashi has also been engaged in clinical ethics consultation as the Founding Director of the Clinical Ethics Center at the University of Tokyo Hospital. A part-time physician, he has worked in hospice care for ten years.

Outside Japan, he has served as a member of UNESCO's International Bioethics Committee and was a board member of the International Association of Bioethics. He is currently a member of the International Advisory Board, Oxford Centre for Practical Ethics, Oxford. Professor Akabayashi serves as an editorial advisor for several international bioethics journals such as *Journal of Medical Ethics, Cambridge Quarterly of Healthcare Ethics, BMC Medical Ethics*, and *Asian Bioethics Review* and is a founding editor of the *CBEL Report*.

Professor Akabayashi established the first Bioethics Center in Asia (University of Tokyo Center of Biomedical Ethics and Law: CBEL, www.cbel.jp) in 2003. This center was acknowledged as a Global Center of Excellence in Japan, highlighting his strong leadership. This leadership is also reflected internationally, through

Professor Akabayashi's establishment of the Global Alliance of Biomedical Ethics Centers (GABEX) that comprises world-class bioethics centers including The Hastings Center, USA, and the Ethox Center, Oxford University, UK. Work done by this network has resulted in a book *The Future of Bioethics: International Dialogues*, edited by Professor Akabayashi (Oxford University Press, Oxford, 2014). His current interest focuses on global bioethics, expanding dialogues among bioethics researchers throughout the world. He has also been honored as a *Fellow of The Hastings Center* (USA) in 2008. He was elected as the *President* of Japan Association for Bioethics in 2017.

Disclaimer: Professor Akabayashi is President of the Japan Association for Bioethics (JAB); however, this book reflects the author's personal academic analyses and opinions. It does not represent JAB's official position.

Chapter 1
A Brief History of Bioethics in Japan

Abstract In this chapter I look back at the history of bioethics in Japan, which can be divided into three phases: Phase I, Introduction (1980–1999); Phase II, Development (2000–2010); and Phase III, the Recent Past (2011–present). Phase I marks the period when the concept of bioethics that originated in the West came to Japan. It was also when Japanese society faced its first difficult bioethical issues: namely brain-death and organ transplantation. Other issues emerged during this period, particularly pertaining to death, such as end-of-life medical care and euthanasia. In Phase II, the problems shifted to those pertaining to the beginning of life, such as the moral status of the human embryo. As well, during this period the government implemented ethical guidelines for research ethics. During this period, social awareness of bioethics increased, and bioethics education began to appear not only medical education, but also within high school curricula. In Phase III, Japan began to tackle its own ethical issues, such as enhancement, regenerative medicine, neuroethics, public health ethics, and precision medicine. Some of my thoughts concerning projections for the future are discussed at the end of this chapter.

In this introductory chapter, I consider the current period (ca. 2020), look back on the history of bioethics in Japan over the past 40 years, and finally, look briefly toward the future. First, I present an overview of the types of problems that have been handled in the fields of bioethics and medical ethics in Japan since the 1980s. I begin my discussion in this era because modern bioethics began its development in Japan in the early 1980s. I divide this time period (beginning in the early 1980s until around 2020) into three parts, corresponding to Phase I, Introduction: 1980–1999; Phase II, Development: 2000–2010; and Phase III: the Recent Past (2011-present). No significant distinction is intended between use of the terms "bioethics" and "medical ethics." For full description, refer to references [1–3]. This chapter is a brief summary of those three papers and some further considerations.

© The Author(s) 2020
A. Akabayashi, *Bioethics Across the Globe*,
https://doi.org/10.1007/978-981-15-3572-7_1

1.1   Phase I: Introduction (1980–1999)

Bioethics is said to have been born in the USA in the 1960s. In the early 1980s in Japan, bioethics literature was introduced and texts were translated from English, primarily at universities and other academic institutions. My own interest in issues such as euthanasia, dying with dignity, disclosing a cancer diagnosis, abortion, and genetic manipulation developed when I was a medical student in the 1970s. I did my best to learn about these subjects, but at the time, Japan had no academic field equipped to handle these types of problems. While attending an exchange event at the Japan-America Student Conference, an American student informed me, "You are interested in a field called bioethics." This was the first time I heard the term "bioethics." In 1979, Beauchamp and Childress published their first edition of the *Principles of Biomedical Ethics*; this was concurrent with bioethics being established as an academic field in the USA.

1.1.1   Brain-Death and Organ Transplantation

In Japan, the issue of brain-death and organ transplantation was highly influential to the development of bioethics (see Chap. 2). Beginning in the 1980s, nationwide debates began as to whether brain-death constituted human death, and whether organ transplantation from a brain-dead body should be permissible. These issues were discussed in medical circles, religious groups, political groups, the media, and among the general population. With the establishment of a commission on brain-death, and the 1997 enactment of the Organ Transplantation Law, organ transplantation from a brain-dead body was finally deemed permissible, with strict conditions. Specifically, the clear and written expression of the intent to donate from the organ donor (15 years or older) as well as the family's consent were both required. Politically, this marked a major milestone. Several positive outcomes were achieved, including discussion on the definition of death and Japanese views on life and death; this set a healthy tone for how bioethical arguments should be handled in Japan from that point forward.

1.1.2   Informed Consent

At about the same time, the idea of informed consent began to emerge as an issue in clinical settings (see Chap. 3). Informed consent in the clinical context entails an act in which "medical caregivers provide sufficient explanation to those with sound judgment capacity and ensure that the patient understands, while the patient then offers consent of their own volition." Today this is considered common protocol, but such was not always the case in Japan. There was even extensive debate on how to translate the English term "informed consent" into Japanese; today, the term "*info-mudo konsento*" is used, but is presented in *katakana*, the phonetic alphabet used for foreign terms and names. In the course of this discussion, those arguing to protect patient autonomy, basing their arguments on self-determination and the right to know, were in a dominant position. Others countered that argument, stating that prioritizing individual autonomy so heavily is unsuited for Japan. However, an explanation of the patient's condition has now been systematically added to the treatment plan at the time of hospital admission and discharge, such that it is covered by health insurance [4]. Informed consent is therefore a common theme that has been integrated into medical care. This flow of events necessitated a change in paternalism among physicians. It also led to the medical record disclosure system based on the Private Information Protection Law in 2003.

1.1.3 Issues with End-of-Life Medical Care and Euthanasia

One other major event during Phase I was the 1991 Tokai University Euthanasia Incident. In this case, a physician, in response to a clear request from the patient's family, administered potassium chloride (KCl) to a terminal cancer patient [5]. He was found guilty of murder and given a suspended sentence by the Yokohama District Court in 1995. The trial was an unprecedented case in which a doctor performed euthanasia. As the court decision presented the conditions that would allow for active euthanasia in the obiter dictum, it was erroneously relayed to other countries that active euthanasia was legally permissible in Japan. Reports of physician-assisted suicide emerged from abroad, and terms such as euthanasia, mercy killing, and dying with dignity were used without clarity. Currently in Japan, the act of a medical caregiver administering muscle relaxants or KCl to a patient, leading to the latter's death, carries with it a high probability that the medical caregiver will be placed on trial for murder. Notably, unlike several other countries, Japan has not legalized active voluntary euthanasia. In addition, with the increased popularity of hospice or palliative care, fewer cases of "withholding treatment" requested ahead of time by patients are legally problematic. However, the issue of "treatment withdrawal" (removal of an artificial respirator) from a terminal patient has yet to be subject to legal resolution (see Chap. 4).

1.2 Phase II: Development (2000–2010)

There are three characteristics of Phase II. First, the topics of discussion shifted from issues pertaining to the end of life (brain-death, end-of-life medical care, and euthanasia) to those dealing with the beginning of life, particularly the moral status of the embryo and problems related to reproductive medicine. Second, numerous policy guidelines and legislation were produced in the fields of life sciences and medical care, paving the way for the establishment of a framework for policy decisions. Third, bioethics came to be socially acknowledged in the fields of medical education and research.

1.2.1 On the Moral Status of the Embryo

In the second phase, topics in medical ethics shifted to issues pertaining to the beginning of life. Among them, the heightened discussion surrounding the moral status of the human embryo became particularly significant. At this time, research on embryonic stem cells (hereafter, ESCs) had become widespread. Human ESCs have the capacity to differentiate into all kind of tissues and all cell types, and there was great hope for these cells in the field of regenerative medicine. However, in order to create human ESCs, one must destroy a fertilized egg (embryo) that has the capacity to become a human individual in the future. If one perceives that human life is present in an embryo, then this type of use of embryos cannot be approved. In 2004, the Council for Science and Technology Policy and the Experts Panel on Bioethics compiled the document "Basic Principles Concerning the Handling of Human Embryos." This document stated that the human embryo is not a person per se, but it is also not an object. The document instead uses the phrase, "sprout of human life." The writers conclude that a human embryo is worthy of respect and requires careful handling (see Chap. 5). Fundamentally, "The Guidelines for Derivation and Utilization of Human Embryonic Stem Cells" and "The Act on Regulation of Human Cloning Techniques" take the same stance as "The Basic Principles Concerning the Handling of Human Embryos." Given that the moral status of the human embryo is a topic of deep religious and political debate in other countries, it seems that Japan has actually handled this issue with relative ease. However, it remains a mystery as to why abortion—another debate that should be begging many questions about the moral status of an individual prior to birth—did not gain as much momentum as that surrounding the human embryo.

1.2.2 Systematization of the Enactment Processes for the Life Sciences and Medical Care

At the end of the twentieth century, numerous governmental ethical guidelines (Ethical guidelines for human genome/gene analysis research, Ethical guidelines for epidemiological research, and Ethical guidelines for clinical research) as well as a number of laws (The Organ Transplant Law, The Act on Regulation of Human Cloning Techniques) came into the fields of life science and medical care. Many of these were drafted by governmental review boards or committees. These drafting sessions came to include medical and legal specialists as well as ethicists, and representatives from the general public such as journalists. By this time, a new framework had been put in place for policy-making in the fields of life sciences and medical care. This was a linear process that began with forming a public committee comprising specialists from many disciplines, drafting the guidelines, implementing the public comment system, and publishing the final results on the internet, followed by media reporting and revisions over several years. This framework reflected the nature and process of the debate surrounding organ transplantation from brain-dead donors.

1.2.3   Ethics Education in Medicine and in Research

By the year 2000, "Medical Ethics" had been added to the curriculum of medical schools nationwide, and faculty members were tasked with teaching these classes. In the early stages, these classes were taught by faculty in forensic medicine, public health, and philosophy/ethics departments as part of general education courses. However, in 2000, the first graduate level Department of Biomedical Ethics was established at the Kyoto University Graduate School of Medicine, and several full-time faculty members were hired. In 2003, the University of Tokyo Graduate School of Medicine also created a Department of Biomedical Ethics. As this process continued to unfold, the "ethics of medicine" education for students in healthcare-related areas became standard. Meanwhile, the ELSI (Ethical, Legal, and Social Implications) programs in the life sciences and medicine were offered a substantial amount of public and private research funding. By the end of Phase II, bioethics had earned a well-respected place in education and research.

1.3   Phase III: The Recent Past (2011–Present)

In what follows I introduce several topics that have emerged or are just emerging in Japan. These are not unique to Japan, but I would like to emphasize those which have become prominent in the Japanese context.

1.3.1   Enhancement

"Enhancement" refers to a specific use of medical technology, that is not been used to treat or prevent disease. It has been defined as an "intervention aimed to improve the human form and function more than is necessary for maintaining health or recovery" [6]. A more straightforward definition is set forth by Kato, a Japanese philosopher, who defines enhancement as "employing medical technology for a purpose other than treatment." Examples of enhancement include the administration of growth hormones for children who are not suffering from growth restriction, "doping" or the use of steroids to enhance muscle strength in sports, and the use of mind-altering drugs such as Ritalin (methylphenidate) and Prozac (fluoxetine hydrochloride) by healthy individuals to enhance their attention (learning capacity) or mood ("smart pills" or "happy pills"). In the future, we may be dealing with designer babies due to genome editing, increased longevity, and enhancement using a brain–machine interface that can operate machinery by connecting a brain to a computer.

Ideas to improve humans are not new, but developments in medical technology now present the possibility of altering the human body, which has led to many debates surrounding the ethical issues involved. Those opposed to enhancement are concerned about its widespread use, advocating that it is unnatural for human beings to "play God," that there are unknown dangers, and that we would lose sacred values such as the importance of weakness and being interlinked with one another. Meanwhile, those whose stance is one of the passive promotion, that is "so long as each individual's freedom to choose is respected we cannot go so far as to prohibit this," are in favor of enhancement, as are those who actively promote it based on utilitarian principles, arguing that enhancement promotes the happiness and enjoyment of human beings, and thus we are obligated to pursue it.

In Japan, rather than either approving or prohibiting all forms of enhancement, each issue is considered individually. Some also feel that as long as society respects individual self-determination and the multidimensionality of values, blanket prohibition of enhancement cannot be enforced, but some restrictions may be warranted. These discussions will help us to grapple with some important issues related to the nature of human beings and society.

1.3.2   Neuroethics

In recent years, a new and rapidly developing field of study has emerged that deals with ethical problems related to neurology or neuroscience and applicable technologies. This field has come to be known as "neuroethics." Safire defines neuroethics as "the examination of what is right and wrong, good and bad about the treatment of, perfection of, or unwelcome invasion of and worrisome manipulation of the human brain" [7]. Two major developments in neuroscience aided in creating momentum for neuroethics. First, the development of brain functional imaging technology such as PET and fMRI made it possible to observe the brain functions of a living human being. This new technology presents the possibility of obtaining an even more diverse array of information, such as lie detection and clarifying an individual's level of consciousness. However, it also opens up possibilities for mind-reading, or the ability to read the condition of another person's spirit. The second major technological development involves selective pharmacological and anatomical intervention/manipulation of neurological processes. For example, it is now possible to control tremors caused by Parkinson's disease with Deep Brain Stimulation. This technology could potentially be applied to patients with other diseases. Moreover, it could also be used for brain enhancement in healthy people. Brainwashing and mind control for military purposes have also come into the debate.

The thought of a brain–machine interface and chimeras of humans and animals being created evokes science fiction-like images of cyborg production, generating fear among people. Some attempts have been made to resolve this fear through two-way communication between scientists and non-scientists, for example, and by improving scientific literacy among the general population and performing risk assessments and encouraging participation from ordinary citizens. The Japanese government is proactive about neuroscience research and has a national brain science project; the Strategic Research Program for Brain Science. While there are no specific topics unique to Japan, ELSI studies of techniques developed in Japan such as decoded neurofeedback are ongoing.

1.3.3   Ethical Issues Surrounding Regenerative Medicine

In 2012, Dr. Shinya Yamanaka created the Nobel Prize-winning human induced pluripotent (iPS) cells at Kyoto University, using technology that allows for the creation of cells with the same differentiation function as human ESCs by incorporating multiple genes into an adult somatic cell. This technology did not require the destruction of a human embryo, and thus generated much hope for cell transplantation medicine that does not involve ethical issues or rejection responses. It was a breakthrough in the field of life sciences as well as the ethical arena. Currently, iPS cell research has moved from the basic research stage to drug development and even clinical applications such as cell transplantation. The Japanese government is actively promoting iPS cell research. However, were the ethical problems truly and completely eliminated by iPS cell technology? Here, I discuss some of the remaining issues.

The first of these is safety. The fact that these cells can differentiate into many cell types means that the possibility of cancerization cannot be ruled out. In this regard, attempts are being made to create iPS cells (using drugs, for example) that lower the number of, if not eliminate the need for, gene recombinants. This is something that future technological advancements will likely be able to overcome.

Next, whether or not it is permissible to create reproductive cells (sperm or egg cells) from iPS cells is an important ethical issue. In research using human cloned embryos, there was some concern that if the cloned embryo was returned to the uterus, then an individual (a cloned human being) would develop. As such, returning a cloned embryo to the uterus was prohibited by the Act on Regulation of Human Cloning Techniques. For iPS cell research, at least in theory, if it is possible to induce differentiation into sperm and egg cells, then it may also become possible to fertilize these. As was the case with the argument surrounding cloning, problems arise concerning the uniqueness of a human individual.

Another developing discussion in this field concerns differentiation into neurons. For example, if a human neuron is grown for research in an animal brain, there is some concern that this animal will develop a human personality. This may present to us a problem related to fusion between humans and animals at the cellular or genomic level. Another challenge is that of stem cell research, for which various uses can be imagined (for example, controlling the direction of differentiation, genetic recombination, and fusion with cells from other species). The problem for researchers then becomes how much consent to obtain from cell donors, and for cell donors, how much they should understand about the future destinations of their cells.

1.3.4   Public Health Ethics

The course of medical ethics in the second half of the twentieth century can be summed up by the phrase "from paternalism to individual self-determination." Part of the background for this was the change in disease patterns. Namely, from infectious diseases to lifestyle diseases, which led to a change in thinking about disease as something that should be prevented at a population level, to that which should be prevented and treated individually. However, in recent years, the dangers of new and re-emerging infectious diseases such as HIV/AIDS, malaria, SARS, and novel forms of influenza have increased. In addition, concern has mounted for bioterrorism-led smallpox outbreaks, or a recurrence of polio, leading us to refocus our attention on the importance of mass prevention. In Japan, the worldwide spread of the H1N1 influenza virus in 2009 is a recent collective memory. This engendered social debate, especially with regard to who would be the first to receive the vaccine.

It is clear that situations are emerging that cannot be well addressed by the kind of bioethics that prioritizes individual self-determination. The opposition and tension between public welfare and individual freedom can be seen in the debates about quarantines and vaccinations for infectious diseases, discussion on handling vaccine distribution, and individual lives being affected by governmental interventions aimed to promote health. Meanwhile, they also create more difficulties than ever before for equal distribution (justice) of benefit and cost. Simply stated, the central arguments concerning these issues pose the following questions: in what situations should individual autonomy and self-determination be restricted, and what degree of intervention for the sake of the common good and paternalistic interventions should be permitted? A new discussion framework for medical ethics—one that is not limited to the conventional patient–caregiver relationship but rather forms the basis for healthcare policy for the various issues plaguing modern-day public health—is in the process of development.

1.3.5   Precision Medicine

Discussion concerning this topic is just beginning in Japan. Due to groundbreaking developments in genomic medicine, the time will come when at the time of our birth (or indeed at conception), we will know our genetic destiny. This raises many serious ethical issues. For example, if it is discovered at birth that a child would develop a severe disease by the age of 10, what sort of care should that child receive? What would we do if we found out that we would develop dementia by the age of 50? Would we be tested early on and take preventive drugs? Adults may have the capacity to make these sorts of decisions, but the ethics of administering these tests to underage individuals is not as clear. Is testing justifiable in children if the intent is to prevent disease or slow its progression? Should genome editing be conducted, ensuring that the genes causing disease are eliminated prior to birth? Alternatively, is this an area into which humans should not intervene? This would change the nature of medicine entirely.

Similar issues are raised by the development of artificial intelligence, as implementation of this technology may change the physician's role markedly. Currently, in an age in which progress has been realized in genomic medicine, precision medicine will likely become one of the most difficult ethical dilemmas for mid-twenty-first century medical care.

1.4 The Future of Bioethics in Japan

Having touched on the history of bioethics in Japan, I will turn to some issues for the future. First, I anticipate that the life sciences and medical technology will continue to progress further, giving hope to many suffering from currently incurable diseases. Some of the developments, such as immune-checkpoint inhibitors, iPS cells, and ESCs, are technological innovations that may help to overcome conventional ethical problems. However, as is evident in issues related to enhancement and neuroethics, forward progress in medical technology that hides within itself great possibilities can, on the one hand, give us great hope for the development of new treatments, while also creating friction between conventional values and ways of living, causing anxiety in society. Moreover, when technology is first introduced, it is difficult to envision sufficiently the influence it will have. How should society handle the "uncertainties" that inevitably and constantly accompany these new technologies? Constructing a framework to ensure that these are addressed effectively is one major challenge for the future of medical ethics in the Japanese context. In addition, even if the arguments in medical ethics shift over to those concerning the beginning of life against the backdrop of new technological developments, issues with terminal medical care and end of life will continue to cause worry among patients and their families, as well as among medical caregivers involved in treatment.

With regard to the many problems that will inevitably emerge, as well as for those that already exist, attempts to address these will likely occur by establishing guidelines or legislation through Japanese governmental committees, as described above. Thus, the framework required to address these issues, while befitting the Japanese context, also comprises an effective method that listens to external voices and adjusts to create harmony in society in order to resolve ethical issues in life sciences and medical care. Yet, I cannot help but assume that the excessive use of guidelines and laws creates a dependency among medical caregivers and research-ers, such that they lose the motivation to face problems head-on and think deeply about them, or when necessary, communicate to the outside world about issues.

In questioning whether a hospital ethics consultation (a system in which an eth-ics committee or its equivalent would offer advice for individual cases) would even function well in this particular climate, I can imagine numerous medical caregivers who would be baffled by having to find answers to difficult ethical issues in the clinical setting. Indeed, the demand not only to accomplish all they do in their busy workday but also to tackle these ethical issues is a tall order. However, we cannot assume simply that because laws and guidelines are in place, all problems will be resolved quickly. If we hope to resolve these ethical issues, we must take a multifac-eted approach that is both policy-based (guidelines are issued, ethics committees are established, and healthcare systems are improved) and education-based (awareness and problem-solving capacities are increased among medical caregivers and researchers in those fields as a result of medical ethics education). This is what I envision for bioethics in Japan.

1.5   Before Moving on to the Main Chapters

In the chapters that follow, there are many passages in which I question the Japanese government's policies in regard to bioethics and research ethics. However, my intent is not to criticize the government, but rather to bring the issues to light and describe them in their rawness through the lens of Japanese culture. If I do not do this, my concern is that international trust and confidence in Japan's research and medical care will deteriorate over time, and the value quality of Japan as a nation will decrease. This is not a lack of patriotism. As noted in the Introduction, the present book was written to achieve a global scale of bioethical dialogue; to this end, I can-not shy away from criticism of my own country's policies or cultures in this process when this is necessary.

The late Japanese scholar Donald Keene, in dialogue with Jakucho Setouchi in the book "*Nihon no bitoku* [The Virtues of Japan]," said, "since obtaining Japanese citizenship, I have come to think that I should …express my opinions as a Japanese citizen," and, "now that I have become a Japanese citizen, I intend to freely speak out against Japan in ways that I have refrained from in the past."

There can be no deep-rooted democracy in a society where the act of criticism itself comes under critique even before validation of the substance of the criticism. In Japanese society today, however, many people internalize an obedient spirit of servility toward current systems, without so much as questioning authority.

I too have been quite hesitant to criticize Japan up to this point, but like Keene, I must say what should be said before I reach the final years of my life. In this book, I speak of what I call "the dark side" of Japan, as seen through the lens of bioethics, so that non-Japanese readers can truly understand Japan. There are already enough publications which describe Japan's advantages.

References

1. Akabayashi A. Bioethics in Japan (1980-2009): importation, development, and the future. Asian Bioethics Rev. 2009;1(3):267–78.
2. Akabayashi A, Slingsby BT. Biomedical ethics in Japan: the second stage. Camb Q Healthc Ethics. 2003;12(3):261–4.
3. Akabayashi A. Nurturing the future of medical ethics in Japan. In: Yasaki Y, editor. Future of medicine. Tokyo: Iwanami-shoten; 2011. p. 55–72. (in Japanese).
4. Akabayashi A, Fetters MD. Paying for informed consent. J Med Ethics. 2000;26(3):212–4.
5. Akabayashi A. Euthanasia, assisted suicide, and cessation of life support: Japan's policy, law, and an analysis of whistle blowing in two recent mercy killing cases. Soc Sci Med. 2002;55:517–27.
6. Juengst ET. Enhancement uses of medical technology. In: Stephen GP, editor. Encyclopedia of bioethics. 3rd ed. New York: Macmillan Reference; 2004. p. 7531–757.
7. Safire W. Introduction. In: Markus SJ, editor. Neuroethics: mapping the field. New York: Dana Foundation; 2002. p. 5.

Chapter 2
Brain-Death and Organ Transplantation: The First Japanese Path

Abstract By analyzing the enactment process of the Organ Transplantation Law (OLT) in Japan, I illustrate one characteristic of the Japanese way to address bioethical issues. The final version of the bill did not establish a blanket definition of brain-death as equivalent to human death. Instead, it suggests that brain-death is the end of life only for patients who have given prior written consent to become organ donors. The family's surrogate consent to donate the organs is not considered sufficient to enact the original 1997 law in every case.

I also examine the extremely low number of brain-dead donors, since the enactment of the Organ Transplant Law twenty years ago, due to, among other things, the Japanese views on corpses (*gotai manzoku*), perspectives of family members, and characteristics of altruism in what I will refer to as the Japanese "village society."

I describe government policy that might support transplant tourism, as well as the background behind the prevalence of living donor organ transplantation. Finally, I refer to publications concerning organ reuse.

At the beginning of each section, I will return to the issue of brain-death and organ transplantation, because this was the first bioethical issue that Japan had to face in the 1960s. In reading about how this issue was handled in Japan, the reader will obtain a better understanding of "The Japanese Path."

2.1 Enactment of the Organ Transplantation Law (OTL)

The movement to legislate organ transplantation from brain-dead persons began in May 1968. However, the first heart transplantation surgery, performed 3 months later (the famous Wada heart transplantation incident), received harsh criticism

© The Author(s) 2020
A. Akabayashi, *Bioethics Across the Globe*,
https://doi.org/10.1007/978-981-15-3572-7_2

from the media, healthcare professionals, and legal experts, because what constituted brain-death diagnosis was not clearly established. There was also suspicion about the status of the donor, and a lack of data around activity on the electro encephalogram. Professor Juro Wada of Sapporo Medical University was charged with a criminal offence (although this ended in non-prosecution). After that the movement toward the legislation for organ transplantation as well as social debate on this issue came to a standstill for some time.

The debate surrounding brain-death and organ transplantation were reignited by the press in the early 1980s, creating a major discussion involving academia, the religious sphere, the government, media, and civil society. Against this backdrop, the development of effective immune-suppressants such as cyclosporine yielded higher success rates in allogeneic organ transplantation surgeries. The debate, which began with the question, "is brain-death the same as human death?" opened up the field for active debate on Japanese perspectives on life, death, and the status of dead bodies, with a large number of publications ensuing. Some of this work was groundbreaking. For example, the clinical or research use of brain-dead bodies "neomort," first introduced by Willard Gaylin in 1974 [1], was further discussed in Japan [2, 3].

On April 24, 1997 in the Lower House of the Japanese Diet, the original bill was voted on and passed, 320 for and 148 against. Then, a revised version passed in the Upper House with 181 for and 62 against on June 17, 1997. That same day, the bill was approved in the Lower House with 323 for and 144 against [4, 5].[1]

In what follows I will focus on the characteristics of Japan's OTL. This law is created through a compromise between those opposed to the notion that brain-death is not affirmed as human death and those who approved of organ transplantation from a brain-dead body. This law does not clearly specify that brain-death is human death. When all of the conditions necessary for organ transplantation from a brain-dead body (that is, the individual's personal consent to organ donation, and family consent) are met, only then is a legal brain-death evaluation conducted, after which organ extraction and transplantation from a brain-dead body is allowed [4].

In Japan, it is common for different ways of being to merge, as is the case with Shinto and Buddhism (syncretism). This is a religious phenomenon in which Shinto (indigenous to Japan) and Buddhism (from abroad) have combined to form a single faith system. Beginning in the Nara era (710–794 A.D.), Shinto gods were celebrated at Buddhist temples and *jin-guji* (places of worship that combined a Buddhist temple with a Shinto shrine) were erected at Shinto shrines.

In this way, Japanese culture often takes two different philosophies or ways of being and fuses them. Thus a resolution is created, but each unique characteristic that is included may also lose some of its intrinsic qualities. Hereafter, I refer to this as a "fusion/transformation" strategy.

Another characteristic of Japan's OTL is the possibility for the family to veto the individual's decision to donate his or her organs. In addition to the potential

---

[1] The road up to that point is described in detail in an excellent book written by the anthropologist Margaret Lock [6], and thus will not be reiterated here.

donor's expressed intent to donate, the law also requires the family's consent. Notably, the 1997 law only considered the expressed intent to donate of those 15 years or older as valid.

2.2   The First Organ Transplant from a Brain-Dead Donor

There was a considerable period of time between the declaration of the first OLT and the first organ transplant. This was partly due to the fact that the 1997 OTL required both a donor's written consent and the family's consent. The family's surrogate consent alone was not sufficient for organ transplantation. It also granted the patient or the patient's family veto power over a diagnosis of brain-death [5]. Thus, the 1997 OTL was one of the strictest "opting in" laws in the world at that time. Many predicted that the strictness of this law would prevent transplantation using organs from brain-dead bodies from becoming a common practice. However, in accordance with the enactment of the OTL in October 1997, The Japan Organ Transplant Network (JOTN) was established, through which registration for heart and liver transplants was initiated, transplant coordinators were appointed, donor cards were made available at many places, and transplant facilities were developed. Nonetheless, Japanese people remained hesitant.

An important Japanese cultural idea that I would like to focus on next is what I will call the "village society." The village society is a form of communitarianism [see 7, 8 for some Western examples], and is a closed and exclusive society within which people are mutually monitored and regulated. In a village society, an act that is considered most challenging is one that disrupts and damages internal homogeneity. For example, if a person were to present an opinion that differs from that of the community, this might be viewed as a betrayal. In this way, individualism in Japan is diluted due to the influence of this traditional form of Japanese culture.

In this form of communitarian culture, if some prominent community members were to be opposed to the idea of organ donation, even though the law had permitted it, becoming the first donor or donor's family would involve some courage.

According to the JOTN, there have been several cases in which potential donors hold valid donor cards, but no organ procurement is performed because the family's consent could not be obtained [9]. Even physicians and organ transplant facilities were also not quite ready to be "the first in Japan," especially knowing that this was a socially contentious issue.

It was against this sociocultural backdrop that the first organ transplant from a brain-dead donor was carried out in Japan [9]. Not surprisingly there was an enormous amount of media attention. Recipient information was supposed to have been kept confidential, and yet detailed information on the donor's sex, age, disease, hospital of admission, and the names of organs transplanted and hospitals where those procedures were performed were disclosed. Essentially, a show was put on for the "village society." This excessive media reporting is thought to have increased anxiety for future donors and their families. It also became a tremendous burden on transplant facilities.

2.3   Twenty years After the 1997 OTL Enactment

As discussed above, the original OTL was extremely rigid, requiring the donor's expressed intent to donate in advance as well as the family's consent. Following its enactment, only 86 cases of organ donations from brain-dead donors took place between 1997 and 2010. In response to these low numbers, the government revised the law in 2010, so that it (1) incorporated an "opt-out" policy that enabled an organ to be extracted with family consent alone if the individual's wishes could not be clarified, and (2) legalized organ transplantation from a child under the age of 15 years with family consent. As a result, the number of organ transplants from brain-dead donors increased under the revised OTL, to 413 cases between 2010 and 2017.

The question remains: how do we interpret this increase? How can we measure its significance? According to the International Registry in Organ Donation and Transplantation (IRODaT) [10], of the 99 registered countries, Japan had an extremely low number of donors in 2015, relative to other countries (Fig. 2.1) [11]. Twenty years have passed since the 1997 enactment of the original OTL, and 9 years since the original OTL was revised. Why have the numbers of brain-dead donors remained so low in Japan?

My colleagues and I performed an in-depth secondary analysis of information published on the JOTN webpage data from the *Fact Book 2016 of Organ Transplantation in Japan* published by the Japan Society for Transplantation [11]. Data from the JOTN webpage comprised results from seven polls conducted by the Cabinet Office between 1998 and 2013 and compared responses to the same questionnaire items. Content analysis was performed to organize the data from 13 families of brain-death donors.

In 2015, livers (8066 from living donors; 318 from brain-dead donors) and kidneys (26,440 from living donors; 649 from brain-dead donors) were the most common organs for transplant. As of June 2017, 615 were registered on the waiting list for a heart, and 318 were registered for a liver. Those registered for a kidney numbered 12,145, and 324,986 were on kidney dialysis.

Secondary analysis of results from polls targeting 3000 individuals from across Japan aged >20 years conducted in 1998 and 2013 revealed some changes in responses over the 15 years under analysis. An increase was noted among those who had declared their intent to donate, especially among younger people. General knowledge about transplantation had increased and the feeling of resistance had decreased. Answers such as "My family will not approve," "I want to leave it up to my family," and "I have no interest" remained steady. In 2013, >90% of the younger group responded that they would respect the family's wishes if they were in writing. However, regarding family consent, those who responded "I will not" [provide family consent] comprised roughly half of all age groups.

As to content analysis, I have selected seven relevant codes extracted out of 13 that relate to the Japanese attitude toward organ transplantation [11]. With regard to family perspective (wherein code names are in { } and typical narratives are in parentheses ()), the codes and narratives are as follows: {family view: to leave it up to the family: *omakase*} (I want to leave this decision up to my family), {For our

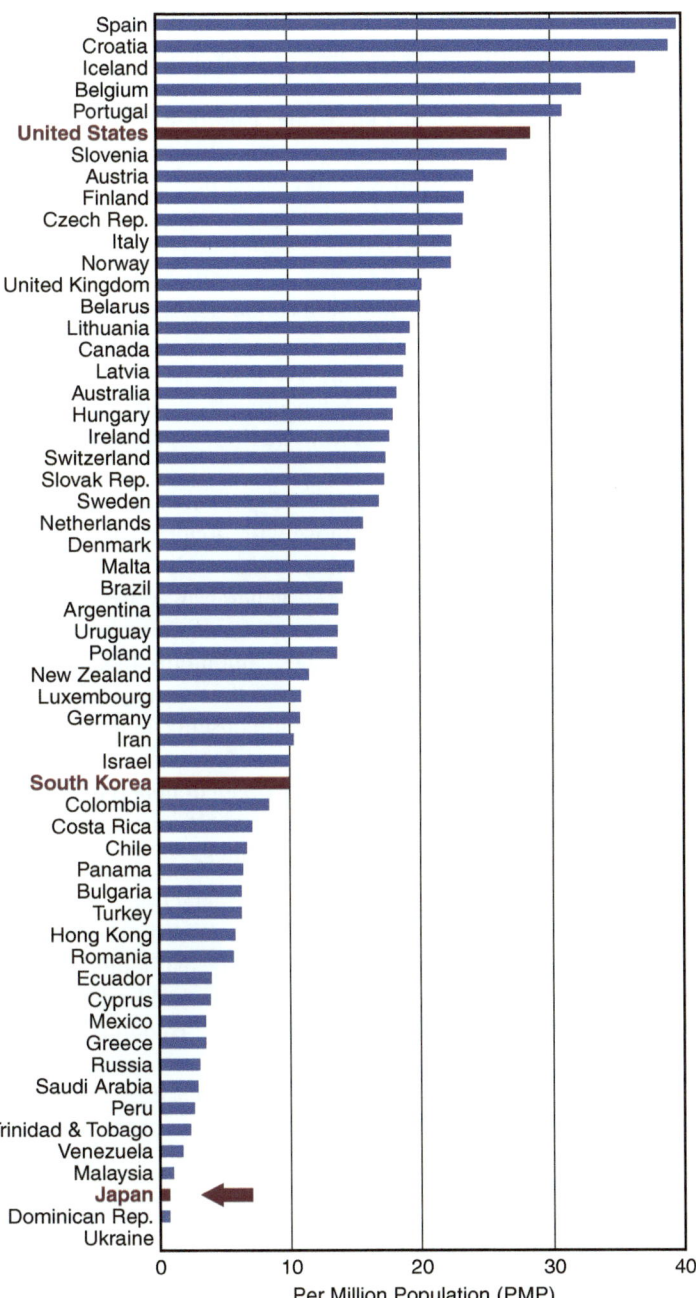

Fig. 2.1 Updated Worldwide Actual Deceased Donors (PMP) in 2015. Data were collected from the 2015 database of the International Registry on Organ Donation and Transplantation. Updated data in 2015 were available for 61 of the 99 registered countries. From among these 61 countries, we selected the same countries listed in the figure of the same database, Worldwide Actual Deceased Donors (PMP) 2013 for the sake of comparison. Four countries are missing data from 2013. Accordingly, data from 57 countries are shown in this figure

own (the family's) sake} (At the very least it might comfort the family), and {Disservice from family/relatives} (I was hurt by insensitive words from those around me). Perspectives from surrogate decision-makers were as follows: {Cannot make the decision} (Even though I know that I need to respect her wishes, I was unable to make that decision with my daughter still alive in front of me), {Worry/ regret afterward} (some donor families still wonder if that was the right thing to do), in addition to {They are living in the recipient} (I can tell the children that part of their mother's body is still living in this world), and {Not my problem, not interested} (just someone distant from me living a very different existence). All of these reflect the tentative attitude in Japan toward (brain-dead) organ transplantation.

Among the issues given as reasons to oppose organ transplants from brain-dead patients during the 1990s, the following five issues seem to have diminished in importance: (1) Distrust in medicine; (2) Uncertainty about brain-death criteria; (3) A lack of infrastructure; (4) Opposition from traditional religions; and (5) Excessive press coverage.

In sum, a comparison of the 1998 and 2013 polls reveal changes over 15 years, including (1) increased proportions of those with donor cards and those declaring their intent to donate; (2) decreased number of those who opt out, due to less resistance toward organ donation; (3) increased general awareness; and (4) increased proportion of those who respect the desires of family members and embrace the concept of "respecting the individual's wishes" among the Japanese people.

Interestingly, in response to the question, "If there is no indication of the individual's intent, would you as a family give your consent?" approximately half of all age groups stated, "I would not."

Notwithstanding, the number of brain-dead organ donors remains extremely low when compared to other countries. Our analysis revealed three sociocultural reasons for this. What follows is the summary of our findings.

1. *Views on corpses* (*gotai manzoku:* "5-body satisfaction," defined as an intact body with a head and four limbs, indicating no defective body parts). This perspective on the body originates from Buddhism. Although Japan is largely a secular country, many Japanese are considered "funeral Buddhists," in that many who would not ascribe to this religion still desire a Buddhist funeral. For this to occur, a corpse must be cremated in a state of *gotai manzoku*. Many Japanese worry that they may not be able to pass into a place of rest without all their organs. Compared to Christian cultures, where the emphasis is more on the soul than the dead body, those who embrace the concept of *gotai manzoku* tend to approve of organ donation far less readily.

2. *Perspectives of the family*. Japan has always been a family oriented society, and strong family bonds persist. In medical care settings, families play important roles in deciding treatment objectives for patients, and patients often leave important decisions up to their families (*omakase*). The patient's body is not just his/her own, but is regarded as one part of a larger whole. Reasons such as {Cannot make judgments}, {Worry/regret afterwards}, or "without *gotai manzoku* I cannot go to the afterlife (cannot be put to rest)" are likely to explain why

some families do not consent to organ transplantation. This trend is not limited to the elderly but is observed consistently throughout all age groups.

3. *Characteristics of Japanese altruism.* The concept of "volunteer" in Japan embodies a slightly different nuance than that in other countries. That is, in Japan, the act of a "volunteer" does not necessarily represent altruism for which no return is expected. While Japan is often viewed as a communal society, in reality, a substantial amount of self-interest is also evident, along with a fairly low level of commitment to others in the community. The average individual in Japanese society has an "inner circle" comprising family and relatives, and all others are considered exactly that: "others." As long as one's inner circle is well, then others do not matter; if there is a problem, it is someone else's to fix. We noted that a certain proportion of individuals responded that they had no interest in organ donation, which may be a byproduct of this form of "altruism."

The views on corpses, families, and this specific type of altruism appear to be deeply rooted. Thus, despite secularization and outward changes in family structure, that which forms the deepest layers of the Japanese spirit persists, crossing generations. I believe that these reasons explain the lack of increase in the number of brain-dead donors, even 20 years after the enactment of the OTL. These numbers will not change easily.

2.4   Is Japan Moving in the Right Direction?

The laws and the necessary environment are already in place in Japan, and yet our cultural perspectives on corpses, on family, and the concept of altruism among Japanese people have not changed. In addition, the requirement of family consent also works against organ donation, which also means that Japanese people's wishes are not being honored posthumously. The lack of cadavic organs has been addressed to some degree through living donor organ transplants, but heart donations cannot come from living donors. The deficiency in hearts available for pediatric transplant is particularly critical.

As of August 2018, only 24 pediatric heart transplants (<15 years old) had been performed in Japan, despite the fact that the 2010 revisions of the Japan's OTL legalized transplants from brain-dead donors under 15 years of age with surrogate consent. The number of overseas pediatric heart transplants to Japanese patients was 67 (5.15 patients per year) between 1998 and 2010, and 31 (4.43 patients per year) between 2010 and 2016 after the 2010 revision of the OTL. There is evidence that nearly all overseas pediatric heart transplants in the USA are Japanese patients. Even after the Istanbul Declaration (2008), the 2010 WHO Global Consultation on transplant tourism, and the 2010 revisions to Japan's OTL, which enabled transplants from child donors under 15 years old, no marked changes have been noted in the yearly number of Japanese pediatric patients undergoing heart transplants overseas.

For this reason, in December 2017, the government targeted pediatric heart patients who could undergo transplants overseas and decided to put in place a policy that would provide partial coverage through the national public health insurance system. On December 22, 2017, the Japanese Ministry of Health, Labor and Welfare (MHLW) issued a notice of a policy in support of transplant tourism. It decided to pay approximately JPY 10,000,000 (about USD 100,000) from the national public health insurance for each patient who underwent transplant surgery overseas. The money could be used toward any purpose except for the purchase of organs, which is prohibited by Japan's OTL.

My colleagues and I performed a critical ethical analysis based upon political philosophy [12]. We found the most promising justification of this policy was to be found in liberal egalitarianism based upon John Rawls' "Difference Principle," in which inequalities are to be addressed to "the greatest benefit of the least advantaged" members of society.

However, it is doubtful that "The Difference Principle," which focuses on the least advantaged, or the "best interests" of (only Japanese) patients, can fully serve as grounds for such justification, to the extent of violating international rules. The issues are as follows: those awaiting transplantation may not be the least advantaged compared to others with chronic conditions. Secondly, this policy infringes upon the opportunity of patients in other countries to access health care and might cause international inequalities surrounding health care access. Therefore, advocating only for the best interests of a limited number of Japanese patients does not offer enough reason to violate international rules.

It would not be strange then, if this policy were to be condemned internationally. Because transplant tourism is, as above, an act which infringes upon the rights to access health care of people from other countries with the same shortages of organs.

Responses vary about Japanese patients undergoing transplants elsewhere. As stated above, we published a critique of this practice in 2018 in a major transplantation journal [12]. There were not many strong responses from Japan. However, outside Japan, there were some strong responses. The Director of The Japan Society of Transplantation replied as follows:

To Members of The Japan Society for Transplantation

December 22, 2017

Director, The Japan Society for Transplantation

This particular system applies to children who will undergo heart transplantations in the USA who were allotted to the "overseas transplantation cases" in order to save lives. ......This is neither organ trade nor industrialism, but rather transplantation at the mercy of US citizens, in which Japanese national health insurance supports Japanese patients who pay the same medical care fees as US citizens. (Translated from the original; author's abbreviations and emphasis)[2]

---

[2] http://www.asas.or.jp/jst/news/doc/info_20171222.pdf.

That said, as the Director of The Japan Society for Transplantation, I feel both deeply apologetic to the pediatric patients in the US awaiting heart transplantations and helpless acknowledging that Japan cannot provide enough pediatric hearts to cover their own transplantation cases.

If Japanese people wish to use organ transplantation, then they have no other alternative but to increase their own numbers of brain-dead donors. Japan is facing a choice: do we value traditional beliefs (perspectives on corpses, familism, our own altruism) or do we invest in a strategy to increase the number of brain-dead donors which would require changes in perspective?

2.5   Living Donor Organ Transplantation

When considering organ transplantation in Japan, we must also consider organ donation from family members. Living organ donation has increased in Japan, perhaps because of the scarcity of organs from brain-dead donors. When I was the Chair of the Ethics Committee of Kyoto University Faculty of Medicine, I interviewed donors and recipients for living organ transplants immediately prior to transplant surgery, in order to obtain final confirmation of their wishes, as a third party not included in the medical team. The interview required them to check the boxes to confirm many items, such as "Were you coerced in any manner by other family members or relatives?" (Table 2.1) [13]. I also confirmed that they knew that they (the donor) could withdraw if they wished to do so, even on the morning of the surgery.

On May 4, 2003, a liver donor died following a procedure performed at Kyoto University Hospital. This was the first official donor death recorded from among approximately 2300 living related liver transplantations performed in Japan. The donor was a mother in her late 40s who donated the right lobe of her liver to her adolescent daughter suffering from liver cirrhosis caused by congenital biliary atresia. After the donation, the liver function in the mother deteriorated rapidly. In January 2003, the mother succumbed to liver failure after an unsuccessful domino liver transplant from a donor with a metabolic disease. She died without regaining consciousness. Histologic examination of the donor liver revealed that the donor had nonalcoholic steatohepatitis (NASH), a rare disease with a poor prognosis [14, 15].

Since then at Kyoto University Hospital, it has become mandatory to confirm that the donor has been informed of the possibility of his or her own death as a worst-case scenario. In fact, as the Chair of the Ethics Committee, I was the one who instructed the medical team about informing the donor candidates that there had been a case in which the donor died. Interestingly, during my interviews with the living donor candidates, they responded that they were indeed aware of this fact; however, the manner in which they responded implied that the likelihood of a donor death was very low (one in several hundred).

Table 2.1 Interview checklist for donors of living related liver transplantation

1. General profile of the recipient and donor
(a) Brief medical history of the recipient
(b) Family tree
2. Informed consent
(a) When and how did you come to know about living related donor liver transplantation?
(b) Who explained the details of the transplant surgery, and how many times?
(c) Under what circumstances (one-to-one, or with others present) were you given an explanation of the details of the transplant surgery?
(d) Do you clearly understand the procedure of the surgery?
(e) Do you fully understand the risks and benefits of the treatment (including the short-term and long-term risks to the donor, and the success rate of graft attachment for the recipient)?
(f) Have you been given information and explanations about alternative therapies?
(g) Have you been given enough time to ask questions? Have you been invited to ask questions?
3. Decision-making process
(a) Have you consulted with anyone on this matter?
(b) Was there any coercion by another family members or relatives? (for example, if you do not agree to be a donor, the patient will surely die?)
(c) Is your decision completely voluntary?
4. Psychosocial aspects
(a) Do you have any anxiety about your surgery?
(b) Do you have any problems in your life (e.g., work or social relationships)?
(c) Do you have any financial problems?
5. Protection of the donor's rights
(a) You have the right to refuse or withdraw your consent until the last moment
(b) You will not suffer any disadvantage if you decide to refuse or withdraw from this
Interview assessment
1. The donor is well informed. ☐Yes ☐No
2. The donor has a good understanding of the entire process. ☐Yes ☐No
3. The donor is fully capable of making a decision. ☐Yes ☐No
4. The donor's decision is completely voluntary and firm. ☐Yes ☐No
5. The decision has been reached without any evidence of coercion. ☐Yes ☐No
6. The donor's right has been fully protected. ☐Yes ☐No
7. The donor is without significant psychological problems. ☐Yes ☐No
Duration of interview   min
Interviewer's signature (a member of ethics committee)
Date

I have heard numerous narratives from donors and recipients of living related organ transplantations during the interviews conducted to confirm informed consent. I noted some hesitancy when donors are asked to check the box that stated that they are not being forced to become a living donor. When I asked them, "Is the intent to donate your liver of your own will?" They all responded, "Yes." However, when they were asked how they came to decide to become a living organ donor the stories became both long and complex.

At least three patterns of informed consent have been developed using qualitative analysis that are relevant here [16]. "*Unconditional Consent*" denotes agreement to donate one's organs without any feelings of reservation; a donor hopes to save a family member's life. Within this category, a donor may occasionally be overzealous, resulting in not hearing accurately issues relating to informed consent. "*Pressured Consent*" describes a donor who feels implicit pressure from others or, from his or her own conscience, to donate. This pressure ultimately overrides any fear the candidate might have toward donation.

"*Ulterior-Motivated Consent*" defines a donor who has an ulterior motive and donates purely for reasons that have nothing to do with unconditional love for one's family. Ulterior-motivated consent is not necessarily monetary, but also relates to psychological reward. For instance, a spouse may offer his or her organ in hopes of patching up a failing marriage.

Another qualitative study using grounded theory revealed the decision-making processes of donors in adult-to-adult living donor liver transplantation [17]. The central theme of this model was "having no choice" and consisted of four codes: (1) priority of life, (2) only living donor liver transplantation, (3) for family, and (4) only me. The model comprised five stages: (1) recognition, (2) digestion, (3) decision-making, (4) reinforcement, and (5) resolution. "Digestion" and "decision-making" described donors' experiences of "reaching a decision;" "reinforcement" and "resolution" stages described those of "facing transplantation." (Fig. 2.2) [17].

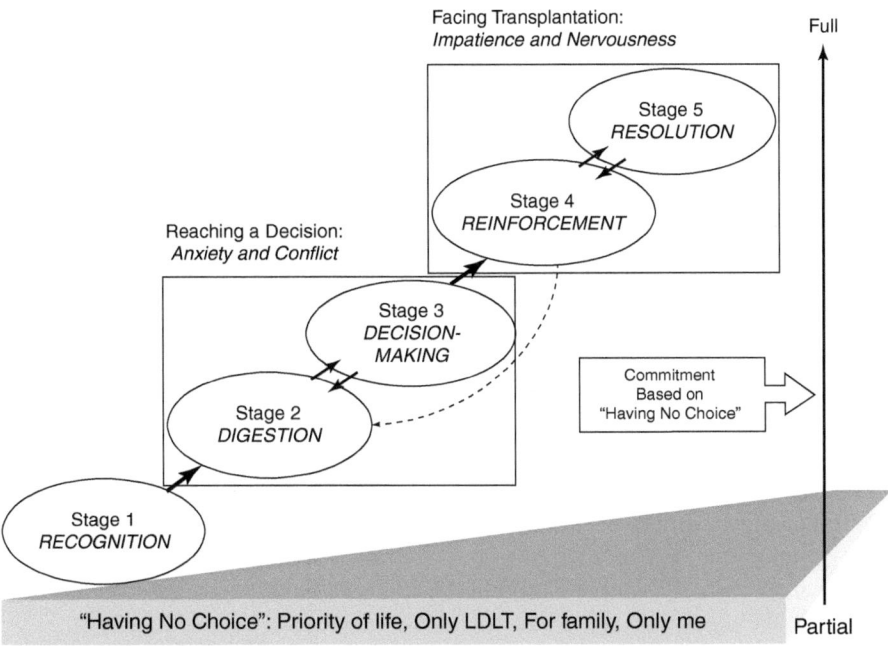

Fig. 2.2 Features of the decision-making model of "having no choice"

Familism, a closed society, and a unique sense of altruism all materialized in various forms through the informed consent process and narratives of each of the donors. Further, living organ donations by anonymous donors reveal the particular characteristics of altruism described above. The Japan Society of Transplantation guidelines do not prohibit anonymous donors. However, one survey reported that the number of members of the Japanese Liver Transplantation Society who would accept friends and strangers as donors was extremely low. By contrast according to a report from the University of Toronto, following the success of the first living liver transplant from an anonymous donor in 2005, over 1000 individuals inquired; of these, 29 were tested and 12 actually donated organs [18]. This would never happen in Japan.

In Japan, where familism is dominant, family donors are relatively easy to come by, and living organ transplantation has developed quickly, partly compensating for small numbers of other categories of donors. National health insurance covers living organ transplantation surgery. For some time to come, living organ transplantation, and not organ transplantation from the brain-dead, will likely continue to be the main transplantation option in Japan. However, the question remains: *"How far should a culture or social value be able to drive a particular medical act, and what ethical issues need to be addressed therein?"*

Addendum Organ Recycling

There is one very unique aspect of how Japanese people view organs. They perceive them as having a sense of inheritability. As the JOTN website states: "They are living in the recipient" and "I can tell her children that part of their mother's body is still living." The emotions contained in these statements are very different from those inherent in the simply utilitarian statement is that, as a person no longer needs their organs, they will pass them on.

In an article on organ reuse [19], my colleagues and I argued that organ recycling endangers the personal identity of the recipient. One issue that was not addressed in that particular article is the perspectives of the donor and their family, not just the perspective of the recipient. We state: "...the organ will inevitably contain its own unique history. This means that the organ will be shouldered with some form of inherent memory, which would form what might be called a 'pseudo personality.'" However, regardless of the body in which the organ lives viewing this issue from the perspectives of the donor and the family rather than the recipient's highlights the fact that it was originally the donor's organ. Thus for the donor family, it will always be a part of someone they loved. In this regard, organs that have been passed on to others do not have a "pseudo personality." For the donor's family, the organ still retains the genuine personality of the donor, and thus some cases of organ recycling do not create major ethical issues. The organ of their loved one is living on in the body of someone else. In fact, knowing that someone else may live on for that much longer may even offer peace of mind to the donor's family with regard to the recycling of this organ.

In response to a paper discussing the same issue in the USA, we presented a hypothetical thought experiment using a case in which a live kidney donor is in a car accident and now requires a kidney [20]. We ask if it is possible for the kidney that was already offered to a recipient to be returned to the donor, and whether this is ethically permissible. The researchers in the USA argue that it is ethically impermissible to impose the risk of dialysis on the recipient. (It is important to note here that in the USA, the prognosis for transplantation is far better than that with dialysis.) They add that the donor with one kidney who donated an organ through living transplantation would be bumped up significantly on the brain-dead transplantation waiting list, and since this individual would likely receive an organ in a short time, there is no need for the kidney that had been donated to be returned to the original donor. However, Japan has excellent results with dialysis and the wait is extremely long for brain-dead transplants. Finally, unlike brain-dead transplants, living organ transplantations are directed donations. Therefore, we argued that it may be permissible to allow one to sign a contract that preemptively specifies that if one were to develop the need for an organ in the future, the kidney recipient could return the organ that was originally donated, instead of becoming a dialysis patient with a severely restricted QOL. The differences between countries this scenario reflects, a prevailing view in Japan that ownership of the organ is still present. This emotion and feeling of inheritability is very common and might also lead to differences in thinking about altruism.

References

1. Gaylin W. Harvesting the dead – the potential for recycling human bodies. Harpers Magazine; 1974. p. 28.
2. Akabayashi A, Morioka M. Research on dead persons. Ann Intern Med. 1989;111(1):89.
3. Akabayashi A, Morioka M. Ethical issues raised by medical use of brain-dead bodies in the 1990s. BioLaw II. 1991;48:S531–8.
4. Akabayashi A. Finally done – Japan's final decision on organ transplantation. Hast Cent Rep. 1997;27:47.
5. Akabayashi A. Japan's parliament passes brain-death law. Lancet. 1997;349:1895.
6. Lock M. Twice dead- organ transplants and the reinvention of death. Berkley: University of California Press; 2002.
7. Sandel M. Liberalism and the limits of justice. Cambridge: Cambridge University Press; 1982.
8. Talyor C. Philosophy and the human sciences: philosophical papers 2. Cambridge: Cambridge University Press; 1985.
9. Akabayashi A. Transplantation from a brain dead donor in Japan. Hast Cent Rep. 1999;29(3):48.
10. International Registry in Organ Donation and Transplantation (IRODaT) 2017. http://www. irodat.org/.
11. Akabayashi A, et al. Twenty years after enactment of the organ transplant law in Japan: why are there still so few deceased donors? Transplant Proc. 2018;50:1209–19.
12. Nakazawa E, Shimanouchi A, Akabayashi A, Akabayashi A. Should the Japanese government support travels for transplantation as a policy under the National Health Insurance system? Transpl Int. 2018;31:670–1.
13. Akabayashi A, Nishimori M, Fujita M, Slingsby BT. Living related liver transplantation: a Japanese experience and development of a checklist for donor's informed consent. Gut. 2003;52:152.

14. Akabayashi A, Slingsby BT, Fujita M. The first donor death after living-related liver transplantation in Japan. Transplantation. 2004;77:634.
15. Neuberger JM, Lucey MR. Analysis and commentary. The first donor death after living-related liver transplantation in Japan. Transplantation. 2004;77:489–90.
16. Fujita M, Slingsby BT, Akabayashi A. Three patterns of voluntary consent in the case of adult to adult living related liver transplantation in Japan. Transplant Proc. 2004;36:1425–8.
17. Fujita M, Akabayashi A, Slingsby BT, Kosugi S, Fujimoto Y, Tanaka K. A model of donors' decision-making in adult-to-adult living donor liver transplantation in Japan: 'having no choice'. Liver Transpl. 2006;12(5):768–74.
18. Reichman TW, et al. Anonymous living liver donation: donor profiles and outcomes. Am J Transplant. 2010;10:2099e104.
19. Nakazawa E, et al. Reuse of cardiac organs in transplantation: an ethical analysis. BMC Med Ethics. 2018;19(1):77. https://doi.org/10.1186/s12910-018-0316-z.
20. Nakazawa E, Yamamoto K, Akabayashi A, Shaw MH, Demme RA, Akabayashi A. Will you give my kidney back? Organ restitution in living-related kidney transplantation: ethical analyses. J Med Ethics. 2020;46(2):144–50. https://doi.org/10.1136/medethics-2019-105507.

Chapter 3
Informed Consent, Familism, and the Nature of Autonomy

Abstract Informed consent is one of the central themes of medical and research ethics. In this chapter, I would like to introduce the reader to three significant cases in the discussion of informed consent from the 1990s, 2000s, and 2010s. Here, I wish to (1) explore the concept of autonomy and the diversity of this term as influenced by the culture and region, (2) explore further the idea of "something close to autonomy" as described by the American bioethicist Edmund Pellegrino in 1992, and (3) articulate to the extent possible the concept of autonomy in Japan. I propose a "family-facilitated approach" to informed consent, which contrasts with the first-person approach used in many Western countries. This family-facilitated model balances respect for patient autonomy with the cultural importance of the family in decision-making, and more clearly characterizes "something close to autonomy" in the Japanese context. I then extend this discussion to the global context.

In the last part of the chapter, I tackle the topic of prognosis disclosure. Although a modern North American concept of autonomy will dictate that physicians inform patients of their prognosis, regardless of patient preference, I argue that disclosing the prognosis at the terminal stage is situation-dependent, and should be decided on a case-by-case basis with consideration of the specific context, based on physicians' virtue, and will open this topic to global dialogue.

3.1 Nature of Informed Consent

This chapter is dependent upon "Informed consent revisited: A global perspective" a chapter in the *Future of Bioethics (2014)* in the Appendix of this book [1]. Of particular significance is the pattern of the three cases described therein, and the places where each occurred [2, 3].

© The Author(s) 2020   27
A. Akabayashi, *Bioethics Across the Globe*,
https://doi.org/10.1007/978-981-15-3572-7_3

Commentaries on this primary article [1] were written by Dr. Carl Becker (Kyoto, Japan, and USA), Dr. Anita Ho (Canada), and Dr. Ruiping Fan (Hong Kong), and face-to-face discussions were held at conferences. Below is a brief summary of the dialogue among the four of us (primarily, my replies to their comments).

Comments to my article can be divided broadly into two categories: First, objections to our appeal to the concept of *autonomy* when supporting the family-facilitated approach. Second, difficulties in practice when implementing the family-facilitated approach.

Let us begin with the first category.

According to Carl Becker, "Dr. Akabayashi's cases are ethically as well as culturally acceptable, not because they somehow approximate patient autonomy, but *because they have unobjectionable outcomes*." He also stated that, "We should respect each culture's way of decision-making even if autonomy is not central to its world view. At the same time, we should seek for ways to avoid *immoral abuse* which cultural systems might permit."

However, it is not clear on what basis Carl Becker can make such evaluative judgments such as *"unobjectionable outcomes"* or *"immoral abuse."* Do these judgments really make sense to those who do not share the same cultural background?

Certainly, someone like Carl Becker, who has sympathy for Japanese culture, would probably intuitively know that the family-facilitated approach in Case 2 was unobjectionable. To others, however, is this really self-evident?

One argument frequently used to defend cultural practices is the claim that there are values specific to a culture, namely values that are only accepted within that culture.

But this argument is problematic. It simply insists on one's own viewpoint and is closed off to criticism from other points of view.

Fan only points out the difference between Confucian moral autonomy and Western conventional autonomy (personal autonomy). He seems to be uninterested in the similarities between these two notions. For example, it might be possible to understand that some elements of Confucian moral autonomy explicated by Joseph Chan—voluntary endorsement, reflective engagement, and importance of the will— share some similarities with conventional Western autonomy.

Our purpose in writing this article was to advance this debate while remaining open to criticisms from others. On this basis we sought to explain and critically scrutinize these practices using concepts and terminology that others can understand, rather than basing our arguments on incommensurable values.

Without looking for such common ground, we are skeptical about whether it would be possible to advance the dialogue between people from different cultural backgrounds.

Thus, a concept used in Japanese decision-making was explained as "a form of autonomy," a notion that is commensurable with other values in other regions of the world. In so doing, I believe that we were able to present a practice specific to Japan in a way that was comprehensible to and open to criticism from those who do not share the same cultural background.

Furthermore, Becker stated that "Dr. Akabayashi's cases provide Western readers a valuable explanation of Japanese decision-making. However, his attempt to defend their similarity to autonomy is questionable." To the contrary, our attempt to defend its similarity to autonomy is by no means questionable.

The primary article clearly articulated the idea of a form of autonomy, namely (1) rejecting paternalism and (2) complying with patient preferences.

Hayashi and I addressed how, in the context of contemporary biomedical ethics, these two points have the potential to apply broadly beyond the bounds of the Japan-USA binary.

Ho also stated that the family-facilitated approach bypasses the substituted judgment standard. In response to this, I want to emphasize the close relationship between the patient and his/her family in the family-facilitated approach.

As Ho implied in her previous article [4], for patients who hold an interdependent view of the self, the family's best interests are equivalent to or at least part of their interests; in these cases, it is impossible to distinguish the patient's interests from those of the family. Thus controversy between the substituted judgment standard and best interest standard is based on the assumption that the patient's interests and those of the family are distinct and potentially incompatible.

By contrast, in the family-facilitated approach, this basic assumption is nullified by the close relationship between patient and family. Consequently, we will set aside the problem concerning the substituted judgment standards. Further, with respect to our notion of autonomy, Ho commented that, "The notion of 'something close to autonomy' or 'a form of autonomy' implies that the family-facilitated approach falls short of being ideal." We do not, however, regard the family-facilitated approach to fall short of being ideal.

Let us recall that "something close to autonomy" or "a form of autonomy" was in fact a quote from the American medical ethicist Pellegrino. From the perspective of the conventional autonomy-centered bioethics approach, the family-facilitated approach may appear non-ideal. Yet, I am not attempting to debate the relative merits of the family-facilitated approach over the first-person approach. That is, we are not saying that the conventional conception of autonomy (personal autonomy by Fan's definition) is better, or worse, than "something close to autonomy."

I will move on to second category: difficulties in practice.

According to Becker, we "need to look less at typically easy cases and more at borderline troubling cases." We believe that each of our three cases represent difficult cases, rather than easy ones.

Perhaps all of the commentators are concerned about the problems that might arise in the application of the idea of tacit consent and the soft proxy approach to actual cases. These concerns are understandable. Nonetheless, our goal in this discussion was to show that, in some cases, the family-facilitated approach is more appropriate than the first-person approach.

We must reaffirm that this argument is constructed based on an ideal situation assuming that two premises have been met ex hypothesi, namely that (1) a patient-family fiduciary relationship exists and (2) a patient identifies her/himself more as a component of the family unit than as an independent individual.

Therefore, the question of what the best response is in a non-ideal situation in which these conditions have not been met, or where it is difficult to know whether the conditions have been met, falls outside our framework. Nonetheless, we would like to respond to these concerns to the best of our abilities.

(a) First, variety of consent as it relates to patient "nodding": In contrast to Becker's understanding, we consider the "nod" in Case 3 as expressed consent rather than tacit consent.

Our argument that there is no difference between tacit consent and expressed consent is aimed at Ho, who stated that there is the danger of exploitation in both cases.

If the two premises are not in effect—in other words, in cases where there is no trust between patient and his/her family—then the hard proxy approach or first-person approaches should represent the appropriate options. We understand that the family-facilitated approach is distinguished from the soft proxy approach in that there is no expressed consent but tacit consent.

(b) Second, a patient's relational identity and trust in her/his family: How do we know?

The first concern with the soft proxy approach, as Becker and Ho point out, relates to the problem of what the physician should do when s/he does not know if there is trust between the patient and his/her family. In other words, how do we verify or know if a patient adopts an independent or interdependent view of the self?

We might hope that in the future, psychological tests will be created. Physicians could then judge whether or not there is trust, or whether a patient holds an independent or interdependent view using this kind of testing.

Without such tests, it is currently difficult to make such judgments. Yet, because there are in reality patients who have an interdependent view of self, the only option is, as Ho suggests, to ask various questions or know a lot more about the patient's background, in order to make inductive inferences and identify such people.

It is possible that Ho interprets the first-person approach as one in which the patient makes decisions about all matters independently. But even in conventional informed consent, the patient is informed by the physician, meaning that the patient is able to decide with the help of the physician. Thus, even according to the first-person approach, the patient requires the assistance of others.

The third concern, which both Becker and Ho raise, is how to handle situations in which the opinions of family members are divided. Indeed, we did not anticipate the possibility of a disagreement within the family. It would be a problem if opinions were seriously divided among family members. In such cases, one may judge that the family lacks decision-making capacity, and thus is unable to decide on behalf of the patient. If available, a call for a clinical ethics consultation may be in order.

Thus, the point made by Becker is extremely astute, as the family-facilitated approach would not be valid. It may be necessary to add a third premise: that the family as a whole possesses the capacity for decision-making.

In such cases, if the patient still desired to have someone else make decisions, the physician might be permitted to decide. This problem requires further study.

That said, I do want to ask Fan to clarify the kind of action the strong family-oriented approach calls for when there are deep divisions in the opinions of family members.

Finally, a characteristic of contemporary medical treatment is that, as the range of treatment options increases, it becomes no longer possible to proceed solely according to a physician's judgment. This is true irrespective of cultural or regional differences. We believe that in this era, it is valid to have a form of autonomy that (1) rejects paternalism and (2) respects the preference of the patient.

In bioethics today, there is a backlash against some kinds of autonomy-centered approaches, which have been challenged by communitarianism and in particular feminist ethics, which conceptualizes autonomy as relational, through which self-hood must be realized through connectedness. Following these different ways of thinking about personhood, we hope that it will be possible to open up new, global developments in bioethical discussion.

Our purpose in writing the article was to advance the debate. Rather than basing the arguments on incommensurable values, we sought to explain our perspective by introducing concepts and terminology for those unfamiliar with Japanese culture. Without finding such common ground, it is doubtful that further dialogue can be sparked between people from different cultural backgrounds. This is what I wrote in the Introduction to this book.

Accordingly, we labeled our central concept *"a form of autonomy"* to explain the Japanese decision-making model, as this is commensurable with values in the West. We believe that this approach enables us to present a Japanese practice that is easily understood by Westerners, while preserving openness to criticism. We also intended to reopen the discussion on informed consent in order to extract ourselves from the dead-end debate over universalism versus ethnocentrism.

Readers are now aware that there are numerous ways of conceptualizing "autonomy," but by using the term "autonomy," researchers from the West can now be cognizant of the fact that there is "a form of autonomy" in Japan, too, and will be able to understand further what this resembles. This, I believe, deepens our mutual understanding. (To obtain a deeper understanding, I recommend those who have access to the *Future of Bioethics* read the commentaries and my replies.)

3.2 Prognosis Disclosure: An Unresolved Issue

In 2003, Japan passed the Private Information Protection Law, which allowed patients to view their own medical charts and learn the name of their disease if they so desired; this demonstrated change around the issue of diagnosis disclosure. However, the next challenge was prognosis disclosure, an issue that remains unresolved worldwide.

According to a 2002 attitude survey ($n = 404$) [5] targeting citizens from the general population in Japan, 86.1% of citizens desired full diagnosis disclosure, while 11.2% desired partial disclosure, demonstrating that nearly all wished for some disclosure. Notably, with regard to expected length of survival (=prognosis), the majority of respondents did not wish for full disclosure, as revealed in the percentages that reportedly desired non-disclosure (5.7%), partial disclosure (47.5%), full gradual disclosure (16.6%), and full disclosure without delay (30.2%).

In a classic attitude survey around the same time (2008) targeting physicians ($n = 710$) in the USA, 98% of physicians informed their patients if an illness would be fatal to them [6]. However, the content of what was communicated differed by respondent. For example, 42% of physicians felt that it is important for patients to know the prognosis and always communicated the prognosis, while 48% of physicians only communicated the prognosis when the patient asked, demonstrating a division into two nearly equal groups. In response to the question, "In telling, how often do you give your patient a specific *timeframe* or medical estimate of the *amount of time* as to when death is likely to occur?", roughly 43% responded "always" or "usually," while 57% responded "sometimes," "rarely," or "never."

What do these numbers indicate? Apparently, even in the USA where respect for first-person oriented autonomy is prioritized so heavily, the practice of prognosis disclosure varies by physician and by patient (Fig. 3.1). Moreover, as predicting a prognosis can be medically ambiguous, the survey results described below are quite thought-provoking.

In a 1999 survey in Japan [7], physicians were asked how a prognosis was conveyed to patients and their families. More than 60% responded that they gave patients a "very" or "somewhat" optimistic prognosis, while more than 40% gave the families a "very" or "somewhat" pessimistic version. Roughly 30% and 60% relayed information solely according to medical judgments to the patients and to the family, respectively.

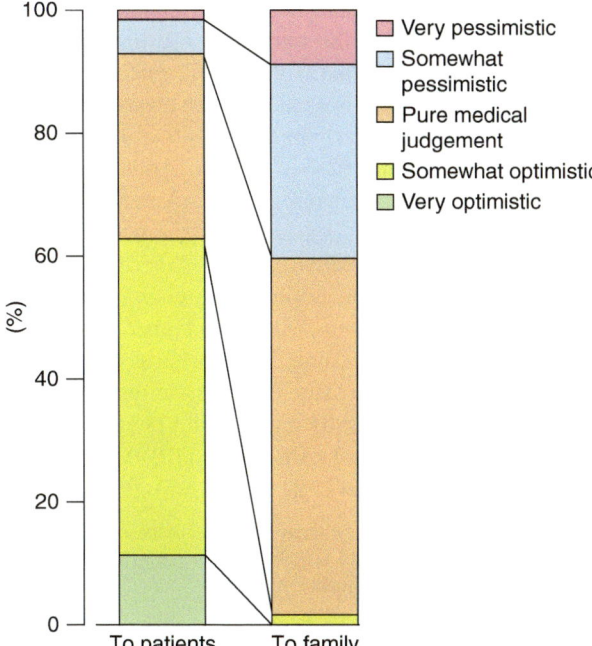

Fig. 3.1 How physicians convey a poor prognosis [7]

When families are offered pessimism while patients are offered optimism about a prognosis, there may be two explanations. Firstly, from the physician's perspective, if they were to show optimism to the family, and then the patient died sooner than expected, they may worry about the patient's family blaming them. Secondly, from the family's perspective, if they receive a pessimistic prognosis and the patient lives longer than expected, this would be a joyful matter. I would describe this interaction as an unconscious psychological conspiracy between the physician and the patient's family. Notably, the patient's own perspective is not included in this.

In an even more recent randomized controlled trial from the USA, physicians delivering the more optimistic message were ranked as more trustworthy. Factors that influence the reluctance of physicians to deliver less optimistic messages to patients with advanced cancer include, among others, fear of being blamed, fear of destroying hope or provoking emotional distress, and fear of confronting their own emotions and death [8].

While the USA also practices a similar type of unconscious psychological conspiracy in disclosing prognosis, a survey from the USA maintains that autonomy is valued, simply because they add the condition, "as long as the patient does not ask." In this way, the same "do not disclose the prognosis" interaction is based on different forms of moral reasoning that vary by culture and region. I hope the readers can recognize the need for sensitivity to differences in ethical grounds used to justify this, even though the act/interaction remains the same. In other words, the same end result is not based on the same moral justification.

I should also note that cultural differences are evident in patient wishes regarding prognosis disclosure. For example, an awareness survey stratified by ethnicity in the USA that targeted elderly individuals found that 35% of Korean-Americans, 48% of Mexican-Americans, 63% of African-Americans, and 69% of European-Americans desired prognosis disclosure. Korean-Americans and Mexican-Americans tended to believe that the family should make the decisions about the use of life support [9] (1995). In addition, while European-Americans and African-Americans felt that the prognosis disclosure is empowering, enabling the patient to make choices, Korean-Americans and Mexican-Americans considered this as "cruel," and "even harmful." The present text does not make cultural comparisons, because as demonstrated above, prognosis disclosure differs so widely in degree, and seems to be handled in a case-by-case manner by patients and physicians all over the world.

A paper was published recently [10] that directly addresses this problem. The authors state the following:

> Does a patient with advanced incurable disease have a right not to hear the bad news? We think not. They should tell the truth, <u>even when the patient would rather not hear it</u>.— (Underlined by the author)

This article presents a questioning of the North American concept of autonomy. Nonetheless, we mount our own objections to this concept. The paper is under preparation, so whether or not our objections are successful remains to be seen. Overall, prognosis disclosure, theoretically and practically, remains the most difficult problem within the realm of end of life communication between patients and healthcare providers.

References

1. Akabayashi A, Hayashi Y. Informed consent revisited: global perspectives. In: Akabayashi A, editor. The future of bioethics: international dialogues. Oxford: Oxford University Press; 2014. p. 735–49.
2. Akabayashi A, Fetters MD, Elwyn TS. Family consent, communication, and advance directive for cancer disclosure: a Japanese case and discussion. J Med Ethics. 1999;25:296–301.
3. Akabayashi A, Slingsby BT. Informed consent revisited: Japan and the U.S. Am J Bioeth. 2006;6(1):9–14.
4. Ho A. Relational autonomy or undue pressure? Family's role in medical decision making. Scand J Caring Sci. 2008;22:128–35.
5. Miyata H, Takahashi M, Saito T, Tachimori H, Kai I. Disclosure preferences regarding cancer diagnosis and prognosis: to tell or not to tell? J Med Ethics. 2005;31(8):447–51.
6. Daugherty CK, Hlubocky FJ. What are terminally ill cancer patients told about their expected deaths? A study of cancer physicians' self-report of prognosis disclosure. J Clin Oncol. 2008;26:5988–93.

7. Akabayashi A, Kai I, Takemura H, Okazaki H. Truth telling in the case of a pessimistic diagnosis in Japan. Lancet. 1999;354:1263.
8. Tanco K, Rhondali W, Perez-Cruz P, Tanzi S, Chisholm GB, Baile W, et al. Patient perception of physician compassion after a more optimistic vs a less optimistic message: a randomized clinical trial. JAMA Oncol. 2015;1(2):176–83. https://doi.org/10.1001/jamaoncol.2014.297.
9. Blackhall LJ, Murphy ST, Frank G, Michel V, Azen S. Ethnicity and attitudes toward patient autonomy. JAMA. 1995;274(10):820–5.
10. Stahl D, Tomlinson T. Is there a right not to know? Nat Rev Clin Oncol. 2017;14(5):259–60. https://doi.org/10.1038/nrclinonc.2017.47.

Chapter 4
End-of-Life Care, Advance Directives, Withholding and Withdrawing Life-Sustaining Treatment, and the Goals of Medicine

Abstract End-of-life care is a universal topic in bioethics throughout the world, with each region, religion, and culture claiming its own position. A comparison of these positions, however, is not the aim of this chapter. Although Japan imported the practice of issuing advance directives (ADs), it has not gained popularity, and I doubt that it will do so in the near future. In addition, from a global perspective, taking into consideration the high infant mortality rates and low adult life expectancy in LMICs, it is fair to assume that AD is not commonplace in those countries. The way in which AD is enacted also depends on local law and culture and is therefore highly contextualized.

In this chapter, I also take up the issue of withholding and withdrawing of life-sustaining treatment, especially the removal of artificial ventilation. There has also been much discussion surrounding the "equivalence principle," which supports withholding and withdrawing. My colleagues and I have challenged this consequentially based idea in an Open Peer Commentary published in the *American Journal of Bioethics* (2019); we wrote this commentary from an East Asian perspective, and I will briefly introduce it.

Finally, in the current era of rapid development in medicine, I have sensed subtle changes in the goals of medicine as perceived by both patients and physicians (Original Article). The emergence of immune-checkpoint inhibitors, for example, may be enough to lead patients to think, "Perhaps if I try to live just a little longer, some dramatic treatment will be developed in the very near future." From comfort care with QOL in the twentieth century to the idea of "a little bit more time in order to be cured" is emerging at the bedside.

No one can avoid death. Ultimately, all physical beings die. Topics in bioethics concerning the end of life include the process of dying and the values—both explicit and implicit—that we hold around this issue. Even actions that may seem to be similar on the surface may have different ethical underpinnings. Discussion around advance directives (ADs) exemplifies this.

4.1   Advance Directives (AD)

In 1998, a survey targeting the general population in Japan yielded some useful descriptive ethical data that helps us to understand the effectiveness of an AD in Japan. Eighty percent of respondents indicated that they would like to express their intents and wishes [1]. What does this mean? Perhaps more importantly, what were the values held by those in the 20% who did not wish to express them?

Issuing an AD may appear to be an expression of unlimited respect for autonomy that ensures that one's own self-determination prevails even after the capacity for decision-making is gone. It also indicates concerns for family. Most respondents who wanted to issue an AD believed that "I want to lessen the burden that my family will shoulder when I'm in the terminal phase of life," indicating that consideration for family was another strong motivator, at least in Japan. Another common response was "because opinions may differ even within my family [1]".

In Japan, the act of issuing an AD seems to incorporate both self-determination as well as concern for one's own family. In other words, respect for autonomy coexists with concern for the family.

The 20% who responded that they would not issue an AD included (in roughly equal proportions) those who noted the theoretical limitations ("I cannot foresee the future") as well as those ascribing from the start to *omakase* ("my family and physician should decide"). Thus, the response that "no AD will be created" represents a coexistence of theoretical limitations and *omakase*.

In response to a question about the level of adherence to the AD, just over 10% responded that their AD should require strict adherence, noting that this was not something for which they would seek a substitute judgment. The general notion surrounding the AD in Japan is that it would ideally serve as a reference against which one's best interest judgment might be determined (that is, the wishes of the individual in question are unclear, so another person must select the best course of action). The question remains: is this consistent with the backdrop against which the AD was developed in the West? Advance directives are technically the same the world over, but the context that gave rise to them, namely respect for autonomy, is not necessarily congruent with the Japanese type of AD.

As of April 2019, roughly 110,000 (0.1% of the population aged 15 years and older) members were registered at the Japan Society for Dying with Dignity (registrants are those leaving a living will). Influenced by the Patient Self-Determination Act of the US in 1990, Japan imported the AD, but it has not gained great popularity. While there are current efforts to promote advance care planning (ACP), implementation of ACP will most likely be due to efforts to uphold self-determination and will also take into account consideration for the family unit. Best interest judgment and the reduction of wasteful medical spending will also be considerations, although careful oversight and governance will be needed in this regard given the possibility for coercion.

4.1.1   AD: A Global Perspective

During the mid-1990s, in a collaborative research setting representing the US, Germany, and Japan, I stated that the AD would be a tool that would become useful, at least to some degree, in diverse cultural settings. Given the difficulties in imagining that AD would be of any use in areas with high infant mortality rates and where health care access and palliative care are both insufficient, my statement [2] pertained to areas of the world in which modern Western medicine was well-established. The likelihood that modern Western medicine would spread and thrive throughout the entire world is low, and even if we were to reach an era and economic state where this was possible, some cultures and religions may be unwilling to accept Western medical paradigms. Therefore, the likelihood that AD and ACP might become universal and global is also low.

4.2   Withholding and Withdrawing Life-Sustaining Treatment (Especially Artificial Ventilation)

Currently in Japan, active euthanasia is illegal, but passive euthanasia, that is, withholding life-sustaining treatment in response to requests by the patient's or the legal surrogate decision-makers is not [3]. Palliative medicine is well-developed in Japan, and palliative care is covered by the national health insurance system. Theoretically, any Japanese person may go to any country where physician-assisted suicide is legal and die there, although there are few reports of Japanese people taking advantage of this option.

The unresolved controversy is that of withdrawing life-sustaining treatment, especially artificial ventilation, from terminally ill patients. This issue has caused some frustration to Japanese patients, families, and healthcare professionals for quite some time.

In 2006, a surgeon withdrew ventilator support from a patient at Imizu Municipal Hospital in Toyama Prefecture, resulting in the patient's death. In response to this, police investigated the case and filed charges, but the case was ultimately dropped due to a lack of evidence. The Ministry of Health, Labour and Welfare (Japan) issued guidelines in 2007 about decision-making procedures for terminal stage patients with no hope of recovery. These guidelines indicate that judgments about withdrawal should be based on the patient's wishes and be made by the medical care team. However, these are abstract guidelines. Moreover, at present (as of December 2019) there is no legal precedent in the Japanese Supreme Court pertaining to the withdrawal of ventilation and thus the fear that a physician might be prosecuted for homicide is legitimate. Furthermore, even if one does not face legal charges, the media in the "village society" may likely impose significant social sanctions on the physician in response to any charges filed by the police.

However, with regard to withholding medical care, there has been no litigation thus far, and as palliative medicine is well developed, negative analysis of this issue by the media has been infrequent. In this way, there are large disparities regarding the awareness of withdrawal and withholding treatment. I would argue that these disparities are caused by differences in legal and cultural interpretations of these two issues.

4.2.1   Legal Perspectives

Some countries consider withholding and withdrawing ventilation from a terminally ill patient as (legally) the same action. Underlying this reasoning is a basic application of the equivalence principle, which allows for both withholding and withdrawing treatment because the result (death) is the same in both cases. This consequentialist-based ideology is prevalent outside of Japan. However, a small portion of legal experts in Japan maintain that withdrawing a ventilator from a terminally ill patient is a "commission" leading directly to the patient's death and thus define this as homicide. The most inhibiting factor is that the Supreme Court has yet to address any case involving the withdrawal of an artificial ventilator from a terminal patient. As stated above, physicians are hesitant to do so, as they fear criminal prosecution. Accordingly, in clinical settings, they continue with futile treatment until the patient's heart stops. This portrays the practical effects of Japan's judicial negligence. Police and the courts take action only after an incident occurs, which delays judicial decision on this matter—in this case, for decades.

4.2.2   Cultural Perspectives

I was surprised to read Ursin's assertion about the situational dependency of judgments pertaining to withholding and withdrawal of treatment at the end of life, which criticized the radical application of the equivalence principle [4]. This article was from the Netherlands. I applaud Ursin's courage in criticizing the overarching and prevailing equivalence principle. In response my colleagues and I wrote an *American Journal of Bioethics* Open Peer Commentary, presenting East Asian perspectives on this matter [5, 6].

In our commentary, we cite the *Jinen hōni* (自然法爾), used often by *Shinran* (1173–263), the founder of the *Jōdō Shinshū* sect of Buddhism. We also refer to the Confucian virtue of *jin* or *ren* (仁) as reasons why the radical equivalence principle is not well accepted in Japan.

The *Jinen hōni* presents the idea that there are no actions committed by humans; rather the world exists in accordance with the laws of nature. The ideology of leaving things to nature carries through to the placing of moral value upon non-action rather than action. In fact, medical professionals and patient families in Japan often use the phrase "leave it to nature."

A similar way of thinking not mentioned in the *AJOB* OPC is well illustrated in *Jūgyūzu* (Ten Ox-Herding Pictures), which consists of ten images and accompanying short poems in Zen iconography that use the herding of an ox as an analogy for training the mind on the path to enlightenment. This was composed by Zen priest *Kakuan* of the *Rinzaishū* sect (a form of Buddhism) during the Běisòng era (960–1127) in China.

> There is no need to decorate or whitewash—just be as you are (*arugamama*). As it is told by the green mountains and blue waters, the beauty of the wide world fills my eyes. Sit quietly, and just behold everything in its *natural* ebb and flow. —(Translated from the original, "*arugamama*." Author's insertion and emphasis)

This "naturalness" or "*arugamama*" is often mentioned by patients and their families as the patient's life is coming to an end, with those involved stating, "Doctor, please let me go as I am (*arugamama*)" or "Doctor, please let her/him go naturally."

As another example of this concept, I would like to introduce the reader to the concept of *jin* (or *ren*), which corresponds to sentiment, the core of which is a feeling of affection. This concept represents a posture of securing, preserving, educating, and nourishing all things as one body with the same root. As such, one's attentiveness would be focused on the patient, regardless of whether they are cognizant of this or not. This sentiment of the family becomes the desire of the family to be with the patient. What connects both the family's sentiment and that of medical professionals is the notion of *jin* (or *ren*); this casts a negative light on withdrawal of medical care at the end of life.

In sum, the fear of prosecution and sanctions by the media and culture are two factors relevant to this issue. In the AJOB OPC, my colleagues and I have argued that decisions to withdraw the ventilator should be made contextually, on a case-by-case basis, and engaging the virtues of the attending physicians. However, in other cultural contexts, I suspect that other reasons may cast doubt on the radical application of the equivalent principle. Cultural diversity also influences judgments pertaining to withholding and withdrawal treatment—another reason why global dialogue is needed in the field of bioethics.

4.3 Subtle Changes in the Goals of Medicine

I was working as a resident of internal medicine in the early 1980s. Dominating our objectives at the time was the acknowledgement of the sanctity of life—otherwise known as the value of enabling a patient to continue living for even one more minute. To this end, my colleagues, supervisors, and I performed cardiac massage for terminal patients with solid cancers when their hearts stopped. From the early

1990s, I served as a part-time hospice physician for more than a decade. I never used an artificial ventilator or epinephrine injections in my hospice care. In the first place, the hospice settings had no ventilators or epinephrine; all I had was a stethoscope, a manometer, and a large supply of morphine ampules. Patients arriving at the hospice room would often comment that "the room looks like a hotel room that I could arrange as I please."

Around the turn of the century, I noticed a subtle change in the bedside "goal of medicine" among both patients and physicians. This change reflected more emphasis on prolonging life, in the hope that new life saving treatments be developed, rather than accepting death.

In a poem by *Ryōkan* (a Buddhist monk in the *Soto* sect from the *Edo* era, 1758–1831), he writes,

> Things which to the end we cannot completely discard
> Even if one abandons all preoccupations, honor, social status, wealth, and family, there are still things which we human beings cannot completely discard. The final self that one should cast off [egotism; to insist upon one's own views and intentions, and not abide by the words of others] is surely the preoccupation with life.—(Translated from the original by the author)

This subtle change in the goal of medicine is already occurring. How to deal with this change will certainly become a central ethical and social challenge by the middle of the twenty-first century. (See Original Article).

Original Article

The Goals of Medicine: Time to Take Another Look

Akira Akabayashi[1,2], MD, PhD; Eisuke Nakazawa[1], PhD; Arthur L. Caplan[2], PhD
[1]Department of Biomedical Ethics, School of Public Health, The University of Tokyo Graduate School of Medicine, Tokyo, Japan
[2]Division of Medical Ethics, Department of Population Health, New York University School of Medicine, New York, NY, USA

Introduction

In May 2018, President Donald Trump signed into federal law the Right-to-Try (RTT). This legislation approved the use of unapproved phase one drugs by patients based upon patient choice, a doctor's certification that death is imminent, and the fact that no other valid treatment options are available [1]. Other nations have also instituted expanded access or compassionate use programs to honor patient requests. In the US, RTT has been criticized for undermining patient safety, while not actually creating access [1–3]. A private system to implement compassionate use requests has been created and performed much better [4]. There is, however, a deeper issue associated with RTT. That is, a subtle change in patient goals in seeking treatment.

Heightened Interest in Compassionate Use (CU)

PubMed was searched for the terms "expanded access," "compassionate use," and "right to try." Fig. 4.1 summarizes changes that have occurred over the past 40 years. A marked increase in the number of CU-related articles published is evident, particularly within the last decade. As PubMed is simply a search engine for publications in the medical field, the increase in this number is not a direct indication of an increase in medical researcher interest. However, an increase in published articles on any clinical subject is evidence that the issue is emerging as significant. A LexisNexis Academic media database search for *New York Times* in 2009 and 2018 showed an increase of articles mentioning "expanded access" from 6 to 42, "compassionate use" from 4 to 16, and the "right to try" 13 to 35. On this basis we suspect that patient requests for RTT or CU are becoming increasingly common.

Dramatic Progress in Medicine Over the Past 20 Years

The progress that has been made in medicine over the past 20 years is significant. Twenty years ago, HIV infection was equivalent to a death sentence. Today, the same disease has become a controllable chronic condition. Hepatitis C can now

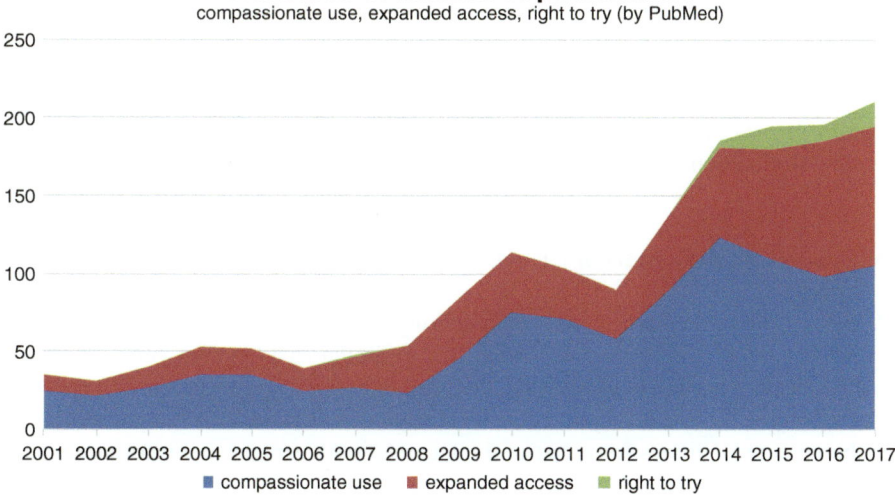

Fig. 4.1 "Expanded access," "compassionate use," and "right to try" were searched across the all fields in PubMed excepting for non-English papers. For "compassionate use," we added the MeSH term. The vertical axis: number of hits. The horizontal axis: year of search

be cured. With the emergence of immune-checkpoint inhibitors [5] such as Nivolumab, many cancer patients who would have until recently been directed to palliative care are now given a realistic extension of functional (high QOL) life expectancy. The evolution of genomic medicine has been dramatic as well [6], and precision medicine is on the brink of major breakthroughs. The time is soon coming when the early prevention of disease onset or controlling the progression of many diseases may well be achieved. In addition, rapid development in regenerative medicine has become a source of hope for patients with diseases with no current viable treatment options, as evidenced, for example, by the pursuit of stem cell remedies.

These facts, as well as extensive media coverage of impending new treatments, evoke a sense of expectation in many patients. For example, some might believe that if they try to live just a little longer, some treatment may be developed in the very near future that could offer a dramatic cure. So a novel goal is emerging—patients are striving to live in the hope of being cured.

The Old Goals of Medicine Are Changing

In 1993, the Hastings Center initiated the Goals of Medicine Project. In the 1996 report, Hanson and Callahan noted four goals of medicine: (1) the prevention of disease and injury and the promotion and maintenance of health; (2) the relief of pain and suffering caused by maladies; (3) the care and cure of those with a malady and the care of those who cannot be cured; and (4) the avoidance of premature death and the pursuit of a peaceful death [7].

With the establishment of the right to self-determination in medical care, beginning in the 1960s, movements to promote dying with dignity and assisted dying gathered momentum. In the 1970s and thereafter, an "increase in QOL" came to be seen as a priority goal for medical care. The central items for medical care in the twentieth century were, therefore, self-determination and QOL. In terms of the goals of medicine, the relief of pain and suffering took a prominent position manifesting, for example, in the growth of palliative care.

Views on Life as Seen from CU and RTL

Views on life in the early twenty-first century have prioritized QOL above all else. However, rapid developments in medical care in the past 10 years brought about a subtle shift—patients now want to "live just a little longer to be cured" and their doctors agree. This contemporary view, "I want to live just a little longer in the hope of being cured," differs from that of the late twentieth century that prioritized QOL, and a peaceful death.

This emerging view represents a synthesis, drawing together highly publicized cutting-edge medical technology. Immune-checkpoint inhibitors and regenerative medicine are both types of medical treatment that enable the simultaneous achievement of high QOL and potentially, cure. These sorts of developments are touted as offering hope.

However, the goal of cure raises tough ethical questions. What odds are worth pursuing? Do marketing and advertising overpromise what is said to be just around the corner? How should patients and their families be supported when the cures do not appear? And how much should any society spend on desperate patients seeking access to unproven treatments? Hope is important, but how far should encouraging hope of an imminent cure guide patient care and public policy?

References (*for Original Article*)
1. The Lancet. False hope with the Right to Try Act. Lancet 2018; 391(10137): 2296. https://doi.org/10.1016/S0140-6736(18)31266-2.
2. Bateman-House A, Kimberly L, Redman B, Dubler N, Caplan AL. Right-to-Try Laws: Hope, Hype, and Unintended Consequences. Annals of Internal Medicine 2015;163: 796–797. https://doi.org/10.7326/M15-0148
3. Caplan AL. Medical Ethicist Arthur Caplan Explains Why He Opposes 'Right-to-Try' Laws. Oncology (Williston Park). 2016 Jan; 30(1): 8.
4. Caplan AL, Teagarden JR, Kearns L, Bateman-House AS, Mitchell E, Arawi T, et al. Fair, just and compassionate: A pilot for making allocation decisions for patients requesting experimental drugs outside of clinical trials Journal of Medical Ethics 2018; 44:761–767.
5. Topalian SL, Hodi FS, Brahmer JR, Gettinger SN, Smith DC, McDermott DF, et al. Safety, activity, and immune correlates of anti-PD-1 antibody in cancer. N Engl J Med. 2012, 366(26):2443–54. https://doi.org/10.1056/NEJMoa1200690.
6. Akabayashi A, Nakazawa E, Caplan AL. Gene editing: Who should decide? Nature 2018; 564: 190. (December 13, 2018).
7. Hanson MJ, Callahan D. Introduction. In The Goals of Medicine: The Forgotten Issues in Health Care Reform. Hanson MJ, Callahan D ed. Georgetown University Press, Washington D.C., 1999; ix–xiv.

References (for Chapter 4)

1. Akabayashi A, Slingsby BT, Kai I. Perspectives on advance directives in Japanese society: a population-based questionnaire survey. BMC Med Ethics. 2003;4(5):E5. http://www.biomed-central.com/1472-6939/4/5
2. Akabayashi A, Voltz R. Advance directives in different cultures. In: Portenoy RK, Bruera E, editors. Topics in palliative care, vol. 5. New York: Oxford University Press; 2001. p. 107–22.

3. Akabayashi A. Euthanasia, assisted suicide, and cessation of life support: Japan's policy, law, and an analysis of whistle blowing in two recent mercy killing cases. Soc Sci Med. 2002;55:517–27.
4. Ursin LØ. Withholding and withdrawing life-sustaining treatment: ethically equivalent? Am J Bioeth. 2019;19(3):10–20. https://doi.org/10.1080/15265161.2018.1561961.
5. Nakazawa E, Yamamoto K, Ozeki-Hayashi R, Akabayashi A. A global dialogue on withholding and withdrawal of medical care: an East Asian perspective. Am J Bioeth. 2019;19(3):50–2. https://doi.org/10.1080/15265161.2018.1563650.
6. Nakazawa E, Yamamoto K, Ozeki-Hayashi R, Akabayashi A. Why can't Japanese people decide?—withdrawal of ventilatory support in end-of-life scenarios and their indecisiveness. Asian Bioethics Rev. 2019;11(4):343–7.

Chapter 5
The Moral Status of the Embryo: The Second Japanese Path

Abstract In this chapter, I illustrate another Japanese strategy for dealing with bioethical issues, in addition to the fusion/transformation strategy that was discussed earlier in the brain-death debate. In order to introduce to the concept of ambiguity in Japan, I refer to the Nobel Lecture by Kenzaburo Oe, who won the Nobel Prize for Literature in 1994.

Following this, in order to explain another Japanese strategy in more detail, I refer to the governmental committee discussion in Japan regarding the moral status of the human embryo following the birth of Dolly, the first cloned sheep, in 1996. This second Japanese strategy was abstracted from the in-depth content analysis of the meeting minutes of the Bioethics Committee of the Council for Science and Technology. In addition to the fusion/transformation strategy discussed around the brain-death debate, the readers will come to understand Japan's "vagueness/ambiguity" strategy which has many advantages but also has significant shortcomings. This strategy is similar to that in which issues are not perceived as black or white and are instead resolved in an indirect way. After all, a human embryo is neither a thing nor a person, but a "sprout of human life." The discussion on the moral status of the human embryo in Japan is, as I explain below, heavily influenced by philosophies of the West.

Lastly, I touch on the discussion of abortion in Japan.

© The Author(s) 2020   47
A. Akabayashi, *Bioethics Across the Globe*,
https://doi.org/10.1007/978-981-15-3572-7_5

In this chapter, I explain how maintaining vagueness/ambiguity is another representative example of the Japanese way of dealing with bioethical issues. First, I refer to a part of the Nobel Lecture by Kenzaburo Oe[1], the 1994 recipient of the Nobel Prize in Literature.

Oe's main objective was to understand Kawabata's stance concerning Japan escaping into its own ambiguous world, where the possibility of foreigners perceiving the culture clearly in unlikely. In addition, Oe considered himself more in line with his Western predecessors in literature who preached the universality of humanism, and less so with his Japanese antecedents.

Let us examine the concepts of *vagueness* and *ambiguity*. Ambiguity shares the same prefix as ambivalent and has is sometimes interpreted as "being in two minds." Oe is thought to have used the term "ambiguity" to emphasize the contradiction of being both "universal" and "closed/inward-looking." However, a close look at the argument in Japan surrounding the moral status of human embryos will reveal more specific knowledge about Japan's form of ambiguity than that revealed through arguments by authors or through cultural interpretation.

5.1   Neither a "Person" nor a "Thing": The Controversy Concerning the Moral and Legal Status of the Human Embryo in Japan

5.1.1   The Bioethics Committee of the Council for Science and Technology

A colleague and I conducted an in-depth content analysis based primarily on records of the proceedings of the Bioethics Committee leading up to the establishment of the Law Concerning Regulations Relating to Human Cloning Techniques and Other

---

[1] Kenzaburo Oe—Nobel Lecture. December 7, 1994. Japan, The Ambiguous, and Myself [1].

Kawabata Yasunari, the first Japanese writer who stood on this platform as a winner of the Nobel Prize for Literature, delivered a lecture entitled Japan, the Beautiful, and Myself. It was at once very beautiful and vague. I have used the English word vague as an equivalent of that word in Japanese aimaina. This Japanese adjective could have several alternatives in English translation.

Kawabata talked about a unique kind of mysticism which is found not only in Japanese thought but also widely Oriental thought. By "unique" I mean here a tendency towards Zen Buddhism. Even as a twentieth-century writer Kawabata depicts his state of mind in terms of the poems written by medieval Zen monks.... According to such poems words are confined within their closed shells. The (non-Zen Buddhism) readers cannot expect that words will ever come out of these poems and get through to us.

I cannot utter in unison with Kawabata the phrase "Japan, the Beautiful and Myself." A moment ago I touched upon the "vagueness" of the title and content of Kawabata's lecture. In the rest of my lecture I would like to use the word "ambiguous."

The modernization of Japan has been orientated toward learning from and imitating the West. Yet Japan is situated in Asia and has firmly maintained its traditional culture......On the other hand, the culture of modern Japan, which implied being thoroughly open to the West or at least that impeded understanding by the West. What was more, Japan was driven into isolation from other Asian countries, not only politically but also socially and culturally (abstracted by the author).

Similar Techniques [2]. Dolly, the cloned sheep, was born in February 1997. In September 1997, the Bioethics Committee was established as a division of the Council for Science and Technology within the Japanese Prime Minister's Office and later renamed the Expert Panel of Bioethics. The Committee comprised 17 members, including life science specialists, medical researchers, clinicians, writers, and researchers in the fields of law, philosophy, religion, and economics. Most of the committee members held positions at the level of Professor or Professor Emeritus. At the beginning committee members were hesitant to make their discussions public, and only a summary of the proceedings was available and the names of individual speakers were not disclosed. However, the committee gradually shifted toward public disclosure, and by the eighth meeting, the committee was completely open, allowing observers to attend.

Initially, the main intent of the Bioethics Committee was to discuss the ethical implications of cloning. To this end, the committee established three subcommittees: the Cloning Subcommittee, the Subcommittee on Human Embryo Research, and the Human Genome Research Subcommittee. Public comments were sought for the reports from each of these committees and were taken into consideration during revisions made by the subcommittees and parent committee. I will pass over the decision of the Cloning Subcommittee, because it was clear, at that point, that human cloning was unanimously prohibited. I also omit the analysis of the Human Genome Research Subcommittee because it is not directly related to this chapter.

I refer primarily to the meeting minutes of the Subcommittee on Human Embryo Research. Through this analysis, the readers will understand how Japan came to an unusual solution to resolve two conflicting Japanese values, establishing that human embryos can be thought of as the "sprout of human life." Below, I extract and summarize parts from the publication [2] which illustrates a unique Japanese way of handling ethical challenges.

5.1.2   The Subcommittee on Human Embryo Research

The draft of the report by the Subcommittee on Human Embryo Research, entitled "Fundamental Policy on Human Embryo Research Focused Primarily on Human Embryonic Stem Cells," was released on March 6, 2000, and states that human embryos are the "sprout of human life" and require respectful, rather than perfunctory, handling. Therefore, research should be permitted within an appropriate regulatory framework. Within the subcommittee it was explained that the concept of "sprout of life" is different from "life" itself, and that research using human embryos should be permitted in certain situations. The subcommittee's report stated the following: [2, pp. 430–431] (authors' translation and abstraction).

> This subcommittee holds that human embryos are the "sprout of human life" and require careful handling, but they are still at a different stage than fetuses or post-natal humans. Therefore, embryos which have been set to be disposed of can be used in research deemed scientifically and morally appropriate... Particularly in light of the establishment of embryonic stem cells....we judge that the use of human embryos should be permitted.

This statement does not contain any moral justification for the use of human embryos. Rather, it focusses on the possibility of scientific use of human embryos as resources for regenerative medicine. The subcommittee suggested the following uses were acceptable. (1) only frozen surplus embryos would be used, (2) consent from the couple would be required, (3) no compensation would be provided, (4) research would be limited to basic research in the first instance, and (5) review by an ethics committee would be required. The emergence of the mysterious term "sprout of human life" will be described in detail in the next section.

5.1.3   *"The Sprout of Human Life"*

The following exchange occurred between committee member Norio Fujisawa, a prominent researcher of Greek philosophy, and legal scholar Ryuichi Ida regarding the expression the "the sprout of human life" [2, pp. 432–433] (authors' translation and abstraction).

Fujisawa noted:

You can rationalize it in all sorts of ways and argue that an embryo is the sprout of life, but not life. Yet, I think that maybe to the ethical sensibility of a normal person, they might be inclined to think that the sprout of life is the same as life."

To which Ida responded as follows:

Well, of course, if born, this becomes a "person." And legally it is possible a fetus would be treated as a person. However, a human embryo has not yet become a fetus.... so I think it is extremely difficult to judge whether this should be thought of as being the same as a person... Having said that, an embryo is not just a "thing" either...it will naturally become a person. This is why we used the expression "sprout of human life," and thus in a sense it describes an intermediate stage on the way to becoming a person... But to hold the position of wanting to conduct human embryo research.... you essentially draw a line to signify that it is acceptable to do certain types of research but not others. In this case, I think that a rationale—that embryos are not human life—is necessary to some degree.

In response, Fujisawa asked:

So then, in terms of ancient philosophical concepts, there's potentiality and actuality. Are you saying that embryos basically constitute the potentiality of life, but not the actuality?

Both Ida and the Imura, the Chair, responded that Fujisawa was correct. However, it is not clear whether Ida and Imura fully understood Fujisawa's question, which was based on the philosophy of Aristotle. However, the exchange suggests that the expression "the sprout of human life" is close to the discussion of "potential," which has been the center of contention in debates over the artificial termination of pregnancy in English-speaking countries. In other words, a human embryo is potentially a person, but it is not yet a person. Therefore, its destruction is not the same as murder [2, p. 433].

The phrase "an intermediate stage" illustrates the subtlety of the Japanese way of thinking. The human embryo is neither a thing nor a person. This report was later accepted by the overarching Bioethics Committee and the committee's views were compiled in a final report, entitled "Regarding Human Embryo Research Focusing Primarily on Human Embryonic Stem Cells." Human cloning was banned by law, while research using human embryos were to be regulated by administrative guidelines.

5.1.4 Consequences

The opposition, comprising scholars and activists, claimed that the Bioethics Committee had not adequately discussed the issue, they should wait for public discussion, the discussion had focused only on short-term usefulness, and that mere lip service was being paid to ethical considerations. However, they also failed to describe what they would define as "adequate" for this discussion.

We concluded the following at the end of our chapter [2, pp. 438–439] (author's abstraction):

> Still the conflict between advocates and skeptics (minority) did not wither away. This conflict was resolved, reaching its climax when the Panel decided to enforce a unilateral decision without consensus in favor of research using human embryos.

> By coining the symbolic phrase "sprout of human life," the Committee found an eclectic solution. Although this term is not a commonsensical one, it nevertheless allowed the Bioethics Committee to approve using human embryos while demanding that researchers exercise due respect in handling them. Accepting the Western idea of human dignity and at the same time making ambiguous the Western dichotomy of things and persons, the Committee made it possible for ethics and science to reconcile with each other, at least for the time being.

In this way, all parties were persuaded, although there was not complete agreement concerning all details. The opposition were (implicitly) satisfied with the appeal made to ethical consideration; namely, the use of the term "the sprout of human life" by the advocates within the Committee. Advocates were satisfied as they had achieved the goal of using human embryos for research, although they could not possibly foresee how those governmental and bureaucratic regulations would ultimately inhibit rather than promote their research later on. (Extremely strict governmental guidelines such as "Guidelines for Derivation and Utilization of Human Embryonic Stem Cells," and "Guidelines on the Handling of Specified Embryos" were issued in 2001 by the Japanese Ministry of Education, Culture, Sports, Science and Technology.)

In this way, when values oppose one another, Japan effectively uses the "vagueness/ambiguity" strategy as a tool to avoid conflict.

5.2   The Issue of Abortion

I have not as yet addressed the issue of abortion. In 1948, Japan enacted the Eugenic Protection Act, and partly due to post-war population adjustments, Japanese citizens felt no strong opposition to abortion. This acceptance of abortion continues into the present. For example, the number of abortions performed in 1954 comprised 50.4% of all pregnancies in Japan. In 2014, reported cases of termination of pregnancy numbered about 182,000, and as total births numbered approximately 1,000,000, it seems that around twenty percent were terminated, not including those terminations that were unreported. Given these circumstances, the question remains: how is the argument that "a human embryo must be handled with respect in research" able to stand?

If we are to treat the human embryo with respect, should abortion even be available? Alternatively, should we treat human embryos with respect, while accepting that abortion is a necessary evil, and also important because of a woman's right to control her own body? While there is a huge literature regarding these issues, reviewing them is not the purpose of this section.

When discussing human embryonic stem cell research, it is doubtful that the current ethical discourse represents the true voice of the Japanese people. Rather it seems more reflective of Judeo-Christian culture. In order to engage with international standards, the Japanese have used vagueness/ambiguity to "fuse" with the values of the West. The result was "the sprout of human life" definition for human embryos, which demanded that researchers exercise due respect in handling them.

It should also be noted that infertile couples undergoing assisted reproductive treatment in clinics in Japan have special emotional connections with their frozen embryos that have not been used for the most recent treatment. These aspects are worthy of further investigation [3].

The concepts of vagueness/ambiguity overlap with fusion/transformation. Thus "the sprout of human life" is a vague and ambiguous term. As is the case for fusion/transformation, vagueness/ambiguity engenders terms that inevitably express ambiguity

Kawabata describes this as vagueness and applauds it as Japanese aestheticism. Oe, on the other hand, labels it as ambiguity and criticizes it as something lacking in universality. Regardless of which term is used, this style of expression undoubtedly forms a core part of Japanese culture.

References

1. Oe K. Nobel Lecture. December 7, 1994. Japan, the Ambiguous, and Myself. https://www.nobelprize.org/nobel_prizes/literature/laureates/1994/oe-lecture.html.
2. Kodama S, Akabayashi A. Neither a "person" nor a "thing": the controversy concerning the moral and legal status of human embryos in Japan. In: Capps BJ, Campbell AV, editors.

Contested cells: global perspectives on the stem cell debate. London: Imperial College Press; 2010.

3. Takahashi S, Fujita M, Fujimoto A, Fujiwara T, Yano T, Tsutsumi O, Taketani Y, Akabayashi A. The decision-making process for the fate of frozen embryos by Japanese infertile women: a qualitative study. BMC Med Ethics. 2012;13:9. https://doi.org/10.1186/1472-6939-13-9.

Chapter 6
The Great East Japan Earthquake and the Fukushima Daiichi Nuclear Power Plant Accident

Abstract The Great East Japan Earthquake on March 11, 2011 caused a massive tsunami that led to the nuclear meltdown at the Fukushima Daiichi Nuclear Power Plant. This horrific accident revealed many systemic flaws, including a weak government program for emergency crisis management, non-transparency within governmental information control, and unscientific approaches to epidemiological research and government funding policy.

Following World War II, Japan prioritized economic recovery over many other things, including preparation for severe natural disasters. My aim is to show how the government handled these emergencies and issues related to research ethics. I will address and criticize the non-transparency of the government's evacuation policy, the secretive position taken by researchers and the government, and the unethical epidemiological research studies conducted under the guise of health surveillance, in particular, child thyroid screening.

I will also discuss the closed nature of the population in Japan's "Village Society." Although Japanese people are known internationally for their courtesy and hospitality, I will discuss the dark side of these traits.

Finally, I discuss environmental ethics, focusing on both animal and intergenerational ethics that were brought to light through the Fukushima accident.

On March 11, 2011, Japan experienced a massive earthquake, magnitude 9.0. As an after effect of the earthquake, the Fukushima Daiichi Nuclear Power Plant was damaged by a tsunami, resulting in a nuclear meltdown. The government's response was by no means laudable, but in fairness the earthquake was of unprecedented severity. Notwithstanding, preparations were insufficient. The reasons for this are based in Japan's prioritization of economic growth after World War II, above all other consideration.

A. Akabayashi, *Bioethics Across the Globe*,
https://doi.org/10.1007/978-981-15-3572-7_6

6.1   Lack of Transparency

One major issue following the earthquake concerned the management of information. Japanese television broadcasting companies, that is all local channels and NHK, the public broadcasting company, continuously broadcast calming imagery, and only made known a small part of the damage. At the same time, the media overseas broadcast images of corpses washed out to sea by the tsunami, or the terrifying conditions of the sites that experienced the earthquake. It was in fact the overseas coverage that created the main impetus for so many to offer international support for restoration and recovery.

The flow of information concerning the status of damage at the Fukushima Daiichi Nuclear Power Plant was infuriating, even to the Japanese media. Detailed, accurate, and real-time information was hidden, particularly by Tokyo Electric Power Company (TEPCO), and residents of Fukushima, who had already been shaken by the disasters were left uninformed, as the area descended further into chaos. There were, in fact, some parallels with what happened 1986 in Chernobyl, but the experiences and lessons learned there were not applied in Fukushima.

The evacuation policy was also poor. After learning of the radiation leak, the government issued an evacuation directive for residents within 20 km of the plant one day after the earthquake. Approximately one month later, this area was designated a 'high alert zone' and effectively sealed off. A colleague and I examined these measures from an ethical perspective and argued that if the government's aim was to avoid health risks posed by radiation exposure, then ordering compulsory expulsion of all residents cannot be ethically justified [1]. It is possible that the government may not have ordered the mandatory evacuation solely based on health risks, but rather to maintain public order. Careful scrutiny of the case revealed that this intervention involved an objective completely unrelated to public health, and that disguising these policies using the purpose of public health made it easier to justify undue restriction of individual liberty.

6.1.1   Closedmindedness, Impenetrability and Secrecy Are Significant Characteristics of Japanese Society

As Oe criticized Kawabata's stance (Chap. 5), likening it to the "flight of Japan to its own vague world, where the possibility for foreigners to gain a correct understanding of Japan is closed off," this impenetrability is evident throughout Japan's history. Japan closed its doors to the rest of the world for over 200 years (1639–1854). During that period, the West made great leaps forward in modernization through the industrial revolution. This "impenetrability" is still present in Japan, even in the age of globalization.

After the Fukushima accident occurred, local residents experienced great distress. However, there was one good that might have emerged: namely, the collection of scientific evidence using empirical and epidemiological methods to measure the (still unclear) effects of low-dose radiation exposure on thyroid cancer development in children. Having observed the confusion between the government and TEPCO immediately after the Fukushima accident, it was my belief that Japan could not singlehandedly conduct such an epidemiological survey. I therefore, through the journal *Science,* called for international collaboration in this research [2].

> Given the current confusion and disorder, it would be difficult for Japanese researchers and the Japanese government to execute such a study singlehandedly. However, they should not have to organize the effort alone. The risk of childhood exposure to radiation is a real one for people living in any region of the world. It is time to organize an international joint research team supported by countries worldwide to uncover lessons to be learned from Fukushima for the sake of future humanity (p. 696).

The response from overseas was overwhelming, and some researchers even offered funding. When invited to serve as a committee member to determine governmental support of the survey of the post-earthquake Tohoku/Fukushima area, I approved governmental support under the following conditions: (1) appropriate relationships are cultivated with residents of Fukushima, (2) sufficient informed consent protocols are conducted, and (3) international cooperation was sought. The principal investigator agreed to all of my stipulations.

However, while foreign researchers were included as advisors, the group in Japan did not seek to make this project an international collaborative study. The most plausible reason for this is that they thought, "What could we gain from these foreigners? They are neither natives of the nuclear disaster-stricken country, nor did they experience the nuclear disaster themselves." This is an example of the closed-mindedness of Japanese society

6.2   The Fukushima Thyroid Screening Study

How much valuable scientific data on low-dose radiation effects have been obtained, or might be obtained through this Fukushima thyroid screening study? In January 2019, a group from Fukushima Medical University (FMU) published the results of the first (2011–2013) and second (2014–2015) rounds of screening for thyroid cancer in *JAMA Otolaryngology-Head & Neck Surgery* [3]. My colleagues and I pointed out several concerns [4].

First, this cohort study was originally designed to obtain scientific data on the effects of low-dose radiation exposure on the thyroid gland in children. Thus, the protocol previously had control groups in Aomori, Yamanashi, and Nagasaki prefectures, which are far from Fukushima prefecture and unaffected by the radiation. However, the sample size ($n = 4,365$) in the control areas was too small to serve as a legitimate comparison to the sample size of those in Fukushima (n=360,000). Therefore, this design has been subject to criticism [5]. Without large-scale controls, the effect of low-dose radiation is difficult to analyze. The FMU group chose to abandon the small control group, and instead used data from the first and second rounds as a baseline. Following the Chernobyl accident, which involved high-dose radiation exposure, the latency of the onset of thyroid cancer was short, roughly 3–4 years after exposure. The estimated latency among people who are iodine sufficient at the time of radiation exposure is thought to be longer, at 5–10 years [6]. Accordingly, the FMU group expected the latency period for the development of thyroid cancer in Fukushima to be 5–10 years, and considered data from the first round (2 years post-disaster) and second round (4 years post-disaster) as the baseline [6]. However, the FMU group paper [3] concluded that 'Large-scale mass US (ultrasound) screening of young people resulted in the diagnosis of a number of thyroid cancers, with no major changes in overall characteristics within 5 years of the 2011 Fukushima nuclear power plant accident,' as if the detected thyroid cancer cases and low-dose radiation exposure were highly unlikely to be related.

There is another serious issue aside from the FMU group using data from the first and second rounds as the baseline. As stated above, if the latency period for the development of thyroid cancer is expected to be 5–10 years, then any effects of the low-dose radiation exposure would begin to show at this time. However, participation rates declined from 81.7% in the first round (2011–2013) and 71.0% in the second round (2014–2015) to 64.6% in the third round (2016–2018). Nonetheless, the FMU authors decided not to show the results from the third round, even though they were available at the time of submission of the manuscript. With this decline in participation rates, precise detection of changes in the onset of thyroid cancer in subsequent rounds is difficult to track.

Among the 202 participants diagnosed with cancer by the second round, more than 80% have undergone surgery. It is highly likely that the participation rates will be much lower for the fourth and fifth round screenings, which will cover 10 years since exposure. This low participation rate is a critical concern.[1]

In maintaining its 'closemindedness,' Japan failed to collect valuable scientific data, potentially the one major contribution to the betterment of humankind that could have been achieved through this disaster.

6.3   Why Less Scientifically Meaningful Data? What About the Victims?

Scientifically speaking, an even more problematic epidemiological survey was undertaken, supported by public funding. Some readers may remember the term, 'Fukushima 50,' which was a label given to the many workers who helped to restore the contaminated site. As they exposed themselves to a highly radioactive environment, they were applauded as heroes. From March 2019, a cohort study targeting

---

[1] In the addendum, I will show an original paper that will explains the decline in participation rates, and inappropriateness of informed consent forms.

those emergency workers was commenced, funded by the Japanese government and conducted by the Radiation Effects Research Foundation based in Hiroshima.

My colleagues and I objected to this study on ethical grounds [7].

Firstly, the low study participation rate is a serious problem. As of March 2018, of the 19,808 workers, 3,400 (17.2%) refused to participate, 7400 (37.4%) did not respond, and 1700 (8.6%) could not be reached. This leaves only 7000 (35.3%) participating workers, most of whom are TEPCO employees. We suspect that the low participation rate may be due to social stigma and fear concerning nuclear power.

Secondly, the unscientific nature of the cohort design further undermines the ethical basis for conducting it. Given the normal statistical variability in cancer incidence and other risk factors, it is unlikely that such increased incidence of cancer due to irradiation would be discernible. The question remains: why did Japanese epidemiologists defend this large-scale cohort study?

We believe that the study should be terminated and the public funding applied instead to activities that truly benefit the workers at the power plant, such as free lifelong health care services and financial compensation.

The Village Society Again: The Case of the Young Woman.
It was my personal experience with a particular young woman that led me to write this chapter.

The story dates back to when I worked part-time at a mental health clinic. To protect the privacy of personal information, the patient's identity and other details are not revealed, but I have tried to present the patient's words exactly as they were spoken. I have also obtained the patient's written informed consent to use her words.

6.3.1   Case

In late 2018, 7 years after the earthquake, I met with a single female patient in her early 30s who had been subjected to harassment in her workplace. She was suffering from insomnia, anxiety, and mild depression. I surmised that she was suffering from an adjustment disorder and began filling out her medical chart accordingly. It was when we got to questions on family composition that I learned that her mother had died, and her father was disabled due to high blood pressure and diabetes. She began to explain that her hometown was in Iwate (Tohoku prefecture). She then shared that the tsunami from the 2011 earthquake not only demolished her entire house, but also washed her mother and a younger female cousin out to sea. Since then, her younger sister had not worked, and instead stayed at home locked up inside, taking care of their father, who is weakened by illness. At this point, my patient began crying uncontrollably. She told me that at her previous workplace, a superior told her, "You are using the earthquake as a crutch." She began to exhibit signs of panic, as I had unintentionally evoked a flashback. This is a typical presentation of PTSD.

My patient told me that she felt her supervisor's statement: "You are using the earthquake as a crutch" was incredibly insensitive. She noted that "when I remember this, I realize that my current workplace is somewhat better than the previous workplace," thus changing her initial complaint.

Hearing the words of her former supervisor made me painfully aware of the reality that Japan's village society possesses a system of cultural stigma in common with most other village societies. To her superior, this woman is not his family, so her emotional state is someone else's problem. The plight of the 2011 earthquake victims was in no way the fault of the victims themselves. However, through this tragedy, the victims came to be differentiated from non-victims, and isolated themselves from the community. This is typical behavior in a village society, in that a portion of the community that has been set apart for whatever reason is discriminated against or ostracized. The government has been more than willing to invest massive amounts of public funding into post-disaster research studies that are merely epidemiological investigations posing as health surveys, but has not reached out to the many victims who still suffer from the after-effects of the earthquake.

6.4   Animal Ethics and Intergenerational Ethics

Finally, let me address animal ethics and intergenerational ethics. As of April 2019, a large area of Fukushima has been designated as a 'high alert zone' and sealed-off. However, animals including pets, livestock and wildlife had all been left in area where there was high radiation. This in it of itself is unethical from the perspective of animal ethics. One research group has examined raccoon teeth to test the presence of a 'dicentric chromosome,' an abnormal chromosome which appears after radiation exposure. In Fukushima, 0.6% of raccoons exhibited this abnormality, while the percentage was 0.0% among raccoons in Aomori (control area, 430 km north of Fukushima).

This finding concerning animal radiation exposure has significant implications for the next generation. Firstly, one imminent issue is that radiation exposure in animals will directly lead to nuclear contamination in surrounding areas. As animals move freely from Fukushima to other regions outside of the high radiation zone, those with high levels of radiation could end up anywhere in Japan. This is problematic for humans and animals alike

Secondly, there is no information on the effects of nuclear radiation exposure on the reproductive systems of small animals. Even larger animals like cows could have reproductive damage that could be passed down to future generations.

By 2020, a wide swath of the area surrounding Fukushima Daiichi Nuclear Power Plant will remain a no-entry zone due to high levels of radiation. Japanese people living today will be leaving the next generation to bear this negative inheritance. Intergenerational ethics is also a globally important issue. Will this culture of "closedmindedness" prioritize "the responsibility for the next generation" or "the values of those of us living now"?

Original Article

Lessons Learned from Fukushima: Thyroid Cancer Screening Preparedness for Radiation Exposure

Akira Akabayashi[1], MD, Eisuke Nakazawa[1], PhD, Hiroyasu Ino[1], MD, and Nancy S. Jecker[2, 3], PhD

[1]Department of Biomedical Ethics, School of Public Health, University of Tokyo Graduate School of Medicine, Tokyo, Japan

[2]Department of Bioethics and Humanities, University of Washington School of Medicine, Seattle, Washington, USA

[3]University of Johannesburg, African Centre for Epistemology and Philosophy of Science, Johannesburg, South Africa

Abstract The 2011 Fukushima nuclear power plant accident prompted much debate among the healthcare sector, especially regarding thyroid gland radiation exposure and follow-up examination. Here, we focus on expertly preparing healthcare systems to address national radiation emergencies, including distinguishing health and epidemiological research, informed consent, and access to services. Drawing on both Japan's experience from the Hiroshima and Nagasaki bombings and its experience in the wake of the Fukushima nuclear accident, we propose key steps for healthcare system readiness. Our proposals will help to improve readiness in the event of future nuclear disasters.

Keywords Fukushima, thyroid screening, epidemiology, health surveillance, informed consent

In the seven years since the 2011 Fukushima nuclear accident in Japan, the academic literature on thyroid screening has grown and public debates within Japan about radiation risk have occurred [1, 2]. Despite greater attention, misunderstandings continue and valuable lessons have yet to be learned and incorporated into policies governing future disaster readiness.

Drawing first on lessons learned from Fukushima we note that the latest Fukushima Report shows 360,000 affected parties and detected 187 malignant or suspicious cases [3]. Among these, more than 50 cases were without sign of invasion or metastasis. At least 11 (over 20%) of 50 participants opted for surgery over the recommended, non-surgical follow-up, based on social and personal reasons, such as parental preference, or no longer residing in the vicinty of Fukushima.

A key report was published offering helpful preparedness advice, especially as it relates to post-accident thyroid cancer screening. The July 2017 EU-OPERA SHAMISEN project report, which sets forth 28 general recommendations, also documents victim fear in the aftermath of radiation accidents [4]. Of the 28 recommendations, the following two pertain specifically to thyroid cancer screening.

R2: Recognise the difference between health/medical surveillance and epidemiology, and their different objectives and data needs.

The objectives of health/medical surveillance are to evaluate whether individuals affected by an accident suffer from some health condition.....In contrast, the objectives of post-accident epidemiology studies are 1) to evaluate whether the radiation exposure/accident has impacted disease rate/risk through "epidemiological surveillance", using population hospital/health-insurance registries; and 2), if possible, to improve our knowledge on effects of radiation, using analytical epidemiological approaches.....

R25: Launch systematic health screening based on appropriate justification and design. Do not recommend systematic thyroid cancer screening, but make it available (with appropriate counselling) to those who request it.

Given the challenge and adverse effects noted above, thyroid cancer screening should be proposed, on a voluntary basis, for those who wish to be monitored, as long as it is accompanied with appropriate information and support.

Here, we reflect on the Fukushima experience and propose healthcare system protocols to expertly prepare healthcare systems for the possibility of future radiation exposures. In particular, we flag ethical considerations related to the distinction between surveillance and research, including informed consent and respect for patient/research subject autonomy.

Epidemiological Surveillance

From the start of the Fukushima study, the boundary between health surveillance and epidemiology research was unclear. Table 6.1 tracks changes in the explanation and informed consent forms provided to participants of three successive studies performed in Fukushima to date.

Table 6.1 Fukushima Medical University's Thyroid Screening Consent Form

During round one (2011–2014), the informed consent document refers to health surveillance [5]. During the second round (2014–2016), the word research appears for the first time in the consent form [6]. The informed consent document explicitly excludes participants from receiving thyroid examinations without consenting to having their data used for research purposes. Epidemiological research thus seems to have commenced as soon as research subjects consented to treatment. During the third round (2016–∗), the informed consent document changes again. This time, it clearly states that the thyroid examination is *not* for the purpose of investigating radiation effects on the thyroid gland [7, 8]; as it appears, the form denies that participants are research subjects in a post-disaster epidemiological study. Yet, at the same time, the round three consent form includes a box to check whether participants 'Agree' or 'Disagree.' This indicates an opt-out format of the third-round. Clearly, the entire process is inconsistent, confusing, and misleading. Whenever a participant opts out of research, it seems they become ineligible for thyroid cancer examination, despite the denial (during round three) that research is ongoing.

Notice regarding Thyroid Gland Examination for the Prefectural Resident Health Survey (2011)

To the guardians of those undergoing the thyroid gland examination:

Because of the radioactive contamination within the prefecture due to the Tokyo Electric Power Company Fukushima nuclear plant accident associated with the Great East Japan Earthquake, Fukushima Prefecture is conducting the Fukushima Prefectural Resident Health Survey, which targets all residents of the prefecture with the aim to secure their safety and relief, and to carry out current and future health management.

As part of the Prefectural Resident Health Survey, in order to carry out the health management of children, Fukushima Medical University (FMU) has been commissioned to conduct thyroid gland examinations that aim to gain an understanding of the current state of their thyroid glands, follow their health throughout life, and provide them and their parents with some peace of mind. The examination is scheduled for October 2011 and will target children from and in the order of Kawamata Town of Yamakita District, Namie Town, and Iitate Village.

The thyroid gland examination will be conducted according to the following operating procedure. Please review the procedure and consider undergoing the examination.

Consent Form for Examination

To Governor, Fukushima Prefecture
To President, Fukushima Medical University

This thyroid gland examination aims to understand the present state of your child 's thyroid gland as part of your child's health management. You will be notified of the examination results, which will also be stored at FMU. This will be done in a manner that ensures privacy, and names will not be traceable. A portion of the examination results might be made public and used as basic information for continuous health management and statistical analysis.

I have read and understand the above, and I provide consent, as guardian, for my
(relation)_____, (name)_____, to undergo thethyroid gland ultrasound examination as part of the Prefectural Resident Health Survey at the locationand date and time indicated on the Notice.

Furthermore, it is noted that my consent is based on the conditions listed below.

                                        Note
(Conditions for consent)
1. I (if an adult, then the person himself/herself) can rescind my consent at any time for any reason.
2. My child(ren) and I will not be disadvantaged in any way if consent is rescinded.
3. Information regarding my child (if an adult, then the person himself/herself) will be provided at any time upon request.
4. Personal information obtained during the present survey about me and my child(ren) will be strictly protected.

Table 6.1 Notices and Consent Sheets used at Fukushima Study from 2011 to 2018

Notice regarding the Second Thyroid Gland Examination (2014)

To those undergoing the thyroid gland examination and guardians:

    Because of the Tokyo Electric Power Company Fukushima nuclear plant accident, Fukushima Prefecture and FMU are conducting thyroid gland examinations to track the long-term health of children.

    To this end, following the first examination (baseline screening), we will be conducting the second (full-scale) examination in order too btain continuous confirmation of the state of the thyroid gland, as described below. We recommend undergoing the examination regardless of the attendance to or results of the first examination.

    Therefore, please review the enclosed "Notice regarding the 2014 Thyroid Gland Examination " and return the necessary information using the self addressed stamped envelope (also enclosed).

Consent Confirmation Form/Medical Questionnaire for the Prefectural Resident Health Survey Thyroid Gland Examination

To Governor, Fukushima Prefecture
To President, Fukushima Medical University

This thyroid gland examination is an examination to track your (the patient himself/herself) health. You will need to submit this consent confirmation form each time you undergo an examination. In addition to notifying you of the examination results, data obtained up until now and in the future in the Prefectural Resident Health Survey will be stored by FMU and used for the following purposes.

- Purposes for your health management
- Management/operation for the survey
- Data provision to external physicians and specialists when opinions and advice are sought
- Collaboration between medical facilities relevant to your examination
- Data provision to municipality in order to offer you appropriate support for health, medical care, welfare, and daily living

- Other purposes
- Use as a basic resource in performing continuous health management for prefectural residents
- Use as a basic resource to further improve and maintain the health survey in the future
- As an educational/training/practicum tool for those conducting the thyroid gland examination
- Use for <u>academic research</u>, public health education or awareness programs, protecting your anonymity
- Use for publication (e.g. statistical analysis), protecting your anonymity

Table 6.1 (continued)

Note
1. You or a legal representative can rescind this consent at any time, and you will not be disadvantaged in any way if consent is rescinded.
2. Information regarding participant will be provided at any time upon request by you or a legal representative.
3. Personal information obtained during the present survey about you and a legal representative will be strictly protected.

Notice regarding the Thyroid Gland Examination (2016)

To those undergoing the thyroid gland examination and guardians:

Because of the Tokyo Electric Power Company Fukushima nuclear plant accident, Fukushima Prefecture and FMU are conducting thyroid gland examinations to track the long-term health of children. This examination aims to observe the state of each individual's thyroid gland in the long-term, to connect this with support that would enable the examinees to live a healthy life, and to aid in a survey regarding future health effects.

This examination will assess the state of the thyroid gland by, e.g., an ultrasound examination, but is not meant to assess the effects of radiation exposure on an individual basis. Since some information regarding the state of the thyroid gland can be obtained from the examination, we will provide you with the results. Although the examination might identify changes that require treatment, which could potentially lead to early detection and treatment, given the characteristics of the thyroid gland, many changes that do not require treatment may also be identified, which may generate some concern. For this very reason, examination of the thyroid gland by ultrasound has not been conducted in general.

Whether or not to undergo the examination depends on the subject's wishes (for those aged 20 years and under, the subject and guardian). Therefore, please consider the contents and significance of the examination and provide us with a reply as to whether you wish to undergo the examination.

Consent Confirmation Form/Medical Questionnaire for the Prefectural Resident Health Survey Thyroid Gland Examination

Same as one used in 2014 except: 'Agree'or 'Disagree'boxes are added.

Updated information in the FMU's Website on May 30, 2016.

In completing the consent form, we have found that some forget to check the box "Agree" or "Disagree." With the exception of those who checked the box "Disagree," we consider even incomplete forms to represent your basic consent to undergo the examination (including the use of data obtained from the testing).

(Authors' translation. Underlined by authors.)

Table 6.1 (continued)

It appears the problems noted above are ongoing. During a 2018 Sectional Meeting of Thyroid Examination Evaluation, a fourth round of the study was discussed, and a new draft informed consent document presented [9]. This draft states, "The aims of this study are to minimize the radiation effects on the thyroid grand,

and to correctly evaluate the relationship between radiation exposure and thyroid cancer." In this way, medical surveillance and post-disaster epidemiological study are conflated.

The participation rate in the Fukushima program has declined from 81.7% in the first round to 70.9% in the second round, and to 54% in the third round [3]. This decline can best be explained by a loss of public trust in the research enterprise, as well as fears of over-diagnosis and overtreatment. Moreover, we suspect that the design barring participants from receiving a thyroid ultrasound examination without consenting to research may contribute to explaining the study's low participation rate. As an alternative, one non-profit organization is offering free ultrasound examinations to parents of children who declined participation in the Fukushima study, but would like their children's thyroids examined [10].

Following the bombings at Nagasaki and Hiroshima in 1945, the Japanese government granted survivors an *Atomic Bomb Survivor's Healthcare Certificate*, guaranteeing life-long free medical services including coverage of a funeral fee [11]. This policy, we believe, facilitated long-term follow-up of survivors' health. The Japanese government also compensated victims, on the premise that governments bear primary responsibility for war and all of its effects. Despite this, the Fukushima study does not cover treatment costs. If victims suffer suspected radiation-related thyroid cancer or leukemia, they themselves must shoulder treatment costs. In a Japanese context, national health insurance system generally covers 70%, and patients are required to pay the remaining 30%. Atomic bomb survivors are exempt from these costs, even if their health problems cannot be shown to originate from radiation exposure.

As of August 6, 2018, the senior management team of Tokyo Electric Power Company (TEPCO) was criminally accused of "professional negligence resulting in death and injury." If the court decides against TEPCO, the company will likely be required to compensate victims for the remainder of their lives, given the magnitude of damages, not only with respect to health, but also quality of daily life, including job and housing losses.

Proposed Health System Protocols

To ensure readiness of healthcare systems for possible future radiation accidents, we propose the following health care system protocols for health surveillance, treatment, and epidemiological study.

Health Surveillance and Treatment Protocols

1. Issue certificates and health notes to all victims to ensure free access to health care services for radiation exposure-related health problems.
2. Conduct health surveillance for victims who consent without linking surveillance to treatment.
3. Clarify evidence-based medical indications for surgical management of thyroid cancer when nodules or cysts are detected by ultrasound examination.

4. Cover treatment costs for radiation exposure-related health problems. Share costs among responsible parties.
5. Assure life-long free treatment for at least thyroid cancer and leukemia, including follow-up for victims outside the disaster-stricken area who remain in Japan. The national government should partly fund follow-up health/medical surveillance to assure nationwide coordination and quality.

Epidemiological Study Protocols

1. Invite child victims together with parents or legal guardians to participate in thyroid cohort research.
2. Use an *opt-out* format for informed consent that assumes consent in the absence of refusal.
3. Distinguish health surveillance and epidemiological research during the recruitment phase. For example, use separate and consistent informed consent sheets.
4. Assure victims during recruitment and treatment that they are eligible to receive thyroid examinations irrespective of whether they opt-in or opt-out of research.
5. Educate participants about the purpose of epidemiological research and emphasize the value of reporting and monitoring precise radiation exposure.

These proposals for improving health system readiness are drawn from Japan's difficult experience of radiation exposure during the Fukushima experiences. Our aim is to help Japan and other nations take necessary steps to become optimally prepared for future radiation disasters.

References (*for Original Article*)
1. Akabayashi A. Fukushima research needs world's support. Science 2011; 333: 696.
2. Normile D. Epidemic of fear. Science 2016; 351: 1022-1023.
3. Yamashita S, Suzuki S, Suzuki S, Shimura H, Saenko V. Lessons from Fukushima: Latest Findings of Thyroid Cancer after the Fukushima Nuclear Power Plant Accident. Thyroid 2018; 28:11-22.
4. Oughton D, Albani V, Barquinero F, et al. Recommendations and procedures for preparedness and health surveillance of populations affected by a radiation accident. SHAMISEN, Nuclear Emergency Situations Improvement of Medical and Health Surveillance, July 2017. https://www.isglobal.org/documents/10179/5808947/SHAMISEN+Recommendations+and+procedures+for+preparedness+and+health+surveillance+of+populations+affected+by+a+radiation+accident+EN/f3df29c3-1c00-4004-91fc-3b0750d5458e (accessed August 5, 2018).
5. https://www.i-repository.net/il/cont/01/G0000338fmu/000/307/000307797.pdf?log=true&mid=JIS-001-20110907&d=1532824232650 (accessed August 5, 2018)
6. https://www.i-repository.net/il/cont/01/G0000338fmu/000/307/000307794.pdf?log=true&mid=JIS-141-20140910&d=1532822644981 (accessed August 5, 2018)

7. http://fukushima-mimamori.jp/thyroid-examination/media/pdf_osirase_dou-isho_download.pdf (accessed August 5, 2018).
8. http://web.archive.org/web/20160530174416/http://fukushima-mimamori.jp:80/thyroid-examination (accessed August 5, 2018).
9. https://www.pref.fukushima.lg.jp/uploaded/attachment/278764.pdf (accessed August 5, 2018)
10. Mother's Radiation Lab Fukushima. https://tarachineiwaki.org/english (accessed August 5, 2018)
11. Supreme Court of Japan, 2007.11.01, Case Number 2005 (Ju) 1977, at: http://www.courts.go.jp/app/hanrei_en/detail?id=916)(accessed August 5, 2018).

References (for Chapter 6)

1. Akabayashi A, Hayashi Y. Mandatory evacuation of residents during the Fukushima nuclear disaster: an ethical analysis. J Public Health (Oxf). 2012;34(3):348–51.
2. Akabayashi A. Fukushima research needs world's support. Science. 2011;333(6043):696.
3. Ohtsuru A, et al. Incidence of thyroid cancer among children and young adults in Fukushima, Japan, screened with 2 rounds of ultrasonography within 5 years of the 2011 Fukushima Daiichi Nuclear Power Station accident. JAMA Otolaryngol Head Neck Surg. 2019;145(1):4–11. https://doi.org/10.1001/jamaoto.2018.3121.
4. Akabayashi A, Nakazawa E, Ino H. Incidence of thyroid cancer among children and young adults in Fukushima, Japan. JAMA Otolaryngol Head Neck Surg. 2019; https://doi.org/10.1001/jamaoto.2019.1105. [Epub ahead of print].
5. Shibuya K, Gilmour S, Oshima A. Time to reconsider thyroid cancer screening in Fukushima. Lancet. 2014;383(9932):1883–4. https://doi.org/10.1016/S0140-6736(14)60909-0.
6. Bauer AJ, Davies L. Why the data from the Fukushima Health Management Survey after the Daiichi Nuclear Power Station accident are important. JAMA Otolaryngol Head Neck Surg. 2019;145(1):11–3. https://doi.org/10.1001/jamaoto.2018.3136.
7. Akabayashi A, Nakazawa E, Ino H, Ozeki-Hayashi R, Jecker NS. Sacrificing the Fukushima 50 again? Journal of Public Health. 2018; https://doi.org/10.1093/pubmed/fdy143. [Epub ahead of print].

Chapter 7
Outcome Egalitarianism and Opportunity Egalitarianism

Abstract In this chapter, I will explore the concept of *equality* in Japan. I ask, what concept of equality is prevalent in Japan? How does this differ from the use of equality in other countries? I will briefly analyze Japan's (1) medical care (universal access to decent minimum fee-for-service care) and the social welfare system, (2) education system (no grade-skipping, teamwork-oriented education), (3) tax and salary system (a progressive taxation system with a high ceiling, a mechanism that strives to ensure that all citizens end up in the middle-income bracket), and (4) some governmental policies (restoration tax at the time of Great East Japan Earthquake in 2011 and dole-out policies in 1999 and 2009). I will explain how "equal burden" and "mutual aids" are key ideas in these social systems using the example of the Japanese concept of equality. I will further describe the system that has been labeled "Japanese socialism."

In summary, I argue here that many of Japan's social systems are based on outcome egalitarianism rather than opportunity egalitarianism.

7.1 Medical Care and the Social Welfare System

7.1.1 Medical Care

In 1961, the national health insurance policy was established for all Japanese citizens, and all were able to enroll in public healthcare insurance. This social policy guarantees all citizens access to medical care. The system employs a fee-for-service model, while remuneration for medical services controls expenditure. In a fee-for-service system, insurance pays the same amount for a procedure, regardless of whether a well-trained surgeon or a new resident performs it. In other words, it is <u>not</u> a system in which the well-trained physician makes the most money. Thus it guarantees access to a decent minimum of medical care, but not to a high quality of

© The Author(s) 2020
A. Akabayashi, *Bioethics Across the Globe*,
https://doi.org/10.1007/978-981-15-3572-7_7

medical care. Financially speaking, the fee-for-service system itself does not reduce costs because it does not prevent over-examination and treatment. However, the Ministry has set relatively low fees, in consultation with the Central Social Insurance Medical Council, an advisory body which represents the public as well as payers and providers. In addition, processes for medical services and drugs are strictly and centrally controlled [1].

In the second half of the twentieth century, when the national health insurance system was established, World War II had recently ended, and medical caregivers worked long hours—unimaginably long by our current standards—and were working endlessly to improve the health of the nation. *"For the patients, for the citizens"* was the mantra of medical care professionalism at the time. If we were to compare the working hours to those of today, Japanese physicians and nurses did not earn a high hourly wage. Surprisingly, this universal health care system, balanced precariously on the public service attitude of medical caregivers, still managed to function.

However, once Japan entered the twenty-first century, the system began to crumble. Due to excessive financial burden, the portion shouldered by patients increased, and universal health care began to change, so that today, medical care has become something that only the rich can afford. Medical caregivers also began to reject excessive work hours. The progress celebrated in medical technology and a sharp escalation in medical fees rendered the system, as it stood, unsustainable.

My aim here is not to provide a detailed account of Japan's medical care system, and thus I will end this passage by emphasizing the "equality" in Japan's medical care: universal access to decent minimum medical care that comprises a fee-for-service system with the fees set relatively low by the government, where well-trained and experienced physicians cannot earn more.

7.1.2 Social Welfare: Public Livelihood Assistance and Pensions

In 2017, roughly 2.14 million persons (1.8% of the total population) were eligible for and received welfare benefits. Of these, 45.5% were 65 years or older. In Japan, those eligible for welfare receive a monthly amount of JPY 137,400 in addition to free medical care and medicines. Thus the living costs alone add up to roughly JPY 3.84 trillion, and such costs are increased if the health benefits for welfare recipients are added.

If a Japanese citizen is employed at a private company between the ages of 20 and 60, a portion of his or her salary is withheld from each paycheck as a pension. As of 2019, a couple who are 65 years and older receive a pension of average JPY 220,000. This is close to an average starting salary of a university graduate. This is another example of outcome egalitarianism. While some discrepancy remains in the amount set aside for each individual, all citizens receive nearly the same amount in their pensions. However, the pension alone is insufficient to cover the cost of living, so those who are able to do so will also prepare by saving for retirement. As regards health care, 10–30% of medical fees for those receiving a pension are covered out-of-pocket.

One may argue that investing heavily in welfare is one form of outcome egalitarianism, in that a nation chooses to assist others so that the outcomes (living standards) are "equal" (with as few disparities as possible). In Japan, this is demonstrated by a general intent "by all citizens to help those in need so that they can all live at roughly the same level as everyone else." The philosophy of outcome egalitarianism—represented in this case by equal burden and mutual aid—plays a major role in this system. This outcome egalitarianism guaranteeing access to medical care and social welfare is rooted not only in medical care, but also in many other policies in Japanese society.

7.2   The Education System

In Japan, the education system is a good measure of outcome equality. It is anticipated that those who enter elementary school in first grade will all move up together to the next class, and that all will graduate together. Teachers and peers help the slow learners. Teachers conduct home visits for troubled students during their free time in order to determine reasons for a high frequency of absences or low grades.

In the period of compulsory education (elementary and junior high schools), students do not skip grades. During these years, individuality is restricted and mutual cooperation is sought from all. Even in higher education, some systems allow for one to skip grades, but these cases are extremely rare in Japan.

Many opinions exist about the ideal form of education. Individuality is important, as is teamwork. In recent years, Japan has been revolutionizing education in an effort to cultivate student individuality and offer a more flexible form of education, among other objectives. However, the ethos in a village society like Japan is teamwork.[1]

7.3   Taxation, Salaries, a Stimulation Policy, and Equal Burden on Individuals

7.3.1   Income Tax and Salary

Japan employs a progressive taxation system in which individuals are taxed according to their income. Until 1988, the maximum ceiling for income tax was 70%.

---

[1] *This section and first part of next section are based on an earlier publication [2].

Even among developed Western nations, this was a very high ceiling. Revisions made in 1988 changed this ceiling to 50%, but even so, the trend in which "the more you make, the more they take" has not changed. The income ratio for the lowest to highest segment of the population was 1 to 2.8 (1994) in Japan, 1 to 8.9 (1985) in the USA, and 1 to 7.0 (1987) in Canada. This mechanism that strives to ensure that all citizens end up in the middle-income bracket is based upon outcome egalitarianism.

Opportunity-based egalitarianism is the opposite. In many Western countries, "equal opportunity" is listed clearly in all job postings. So long as the starting point is the same, any inequality in outcome is acceptable. Therefore, regardless of the huge difference in pay between the CEOs and regular employees of a company, this is not considered "unequal."

7.3.2 Restoration Tax Following the Great East Japan Earthquake and Fukushima Nuclear Power Plant Accident

Following the earthquake and the Fukushima nuclear accident, the government increased income tax by 2.1% and implemented a restoration tax increase over 25 years as a way to help with recovery. The equal burden and mutual aid components of this measure are justifiable because this restoration tax increase is to be used for the restoration of citizens who suffered from natural disaster, so that the victims do not have to shoulder the costs themselves.

Ironically enough, a good portion of the restoration tax (JPY 19 trillion over the 5 years beginning in FY 2011) was reportedly spent in regions outside of those affected by the disaster. For example, some of this tax funded the repair of national highways in Okinawa (far from Tohoku/Fukushima area and unaffected by the earthquake), measures against anti-whaling groups, and restoration of national sports stadiums. The government's underlying rationale for this was—apparently—that (equal) restoration is needed across Japan. However, the restoration tax increase was a policy to enable an equal burden increase by all citizens to help earthquake victims and not to restore Japan overall.

7.3.3 A Stimulation Policy

The government frequently implements an "equal" stimulation policy wherein all citizens become targets for a stimulus package. Many of these have been political in nature, but Japanese citizens have a particular fondness for the term "equality." In addition, most citizens do not distinguish between "opportunity" and "outcome" equality.

Example 1: Regional Promotion Tickets (1999)

Regional promotion tickets were issued by all municipalities in Japan and were worth JPY 20,000 for persons fulfilling the criteria below. In total they cost the Japanese government JPY 619.4 billion. Individuals fulfilling the following conditions as of January 1, 1999 received this funding:

- A head of household with children aged 15 years or younger
- Any recipient of the Old Age Welfare Pension (*rōreifukushi nenkin*), Basic Disability Pension (*shōgai kiso nenkin*), Basic Survivors Pension (*izoku kiso nenkin*), Mother and Child Pension (*boshi nenkin*), Quasi-Mother and Child Pension (*jun boshi nenkin*), Full Orphan's Pension (*iji nenkin*), Child Rearing Allowance (*jidō fuyō teate*), Benefits for Children with Disabilities (*shōgaiji fukushi teate*), and Special Disability Allowance (*tokubetsu shōgaisha teate*).
- Recipients of livelihood subsidies and those placed in a social welfare facility
- Individuals aged 65 years or older who were exempt from municipal taxation

Example 2: Supplementary Income Payments (2009)

Supplementary income payments (*teigaku kyūfu kin*) represent one measure employed as part of the emergency economic stimulus package that was enacted in March 2009, and represented a flat-rate tax reduction policy in the form of supplementary payments.

Every Japanese resident received JPY 12,000. Those who were 65 years and older and those 18 years and under received an additional JPY 8000.

Neither of these two interventions had significant economic effect.

7.4 Japanese Socialism

"Japanese socialism" is a term derived from the assessment that the Japanese economy following World War II "possessed socialist components." The term was used by economists such as Yasuo Takeuchi.

On the one hand, this term holds positive nuance in that it implies that Japanese socialism differs from that of the Soviet Union, in its upholding of freedom and democracy, while creating a communal society in which social disparities are few. However, it also has some negative connotations, such as the fact that the government conducts excessive economic restrictions and interventions. It is often used to criticize the reality that socialist nations have larger disparities and inequalities than those in Japan. It is used ironically to label Japan as the most successful of all communist countries in the world.

Nonetheless, Japan is a democracy. For the 75 years following World War II, through interactions with the USA, including the creation of the Japanese constitution, Japan has been able to avoid direct involvement in any wars that have subsequently developed. Economically, it has become one of the world's economic powerhouses. Despite these achievements I speculate that a good number of

Japanese people are dissatisfied with the current structure based upon outcome egalitarianism. With the massive emigration of many specialists and trained professionals, Japan is in danger of losing strength. Among other laudable qualities of our government are freedoms of speech and academia, both of which allow me to write these words here without fear of imprisonment. Of course, as the village society structure continues to be upheld, I assume that sanctions from some parties will be felt at some point. For now, I am grateful for the current climate in Japan in which I can publish this book, even while knowing that some will interpret my words as criticism.

The concept of equality varies by culture and society. This chapter will teach us that we must not only understand, but also become sensitive to the fact that differences exist not only in how equality is conceptualized by region, but also in the values considered important therein.

Do either of these two forms of egalitarianism function in the global setting? Do equal opportunities actually present themselves to children born in LMICs as readily as they do to children born in developed nations? Is it realistic to strive for economic outcome equality across the world? What kind of equality does UNESCO have in mind when they advocate for this? Do representatives of Africa, for example, wholeheartedly proclaim that they have achieved outcome equality? Is this equal opportunity, outcome equality, or both? Can the two theoretically coexist? It is highly unlikely that we can form consensus about distributive justice at this time.

References

1. Akabayashi A. Financing health care – a Japanese perspective. Health Care Anal. 1995;3:123–5.
2. Ohi G, Akabayashi A, Miyasaka M. Japan's health care system as egalitarian: a brief analysis of its history and selected social systems. Health Care Anal. 1998;6(2):141–9.

Chapter 8
Research Regulations, Ethics Committees, and Confronting Global Standards

Abstract Japan's modern system of scientific governance was imported from the West. Starting with the The International Council for Harmonisation of Technical Requirements for Pharmaceuticals for Human Use: Good clinical practice (ICH-GCP).

Japan needed to institute best global practice in order for Japanese research and pharmaceutical products to be recognized on the world stage. However, because the scientific governance system was imported, there was a mismatch between the basic concepts and underlying intentions of the system and Japanese culture.

The lack of competent ethics committee members and ethics consultation support infrastructure in Japan is a serious concern. Scientific misconduct is a universal phenomenon; however, I will explore it here in the local context. Finally, I will illustrate how the advice by clinical ethics consultations differs from culture to culture, although the formats (individuals, teams, and committees) are the same.

Japan imported research regulations from the West, driven by concern that if it did not meet international standards, then its medical research and drug development would not be respected. The earliest established protocol was the Good Clinical Practice guidelines (GCP, 1989) that regulated drug trials. Japan revised its standards to be consistent with those of the GCP (ICH-GCP) issued at the International Conference on Harmonization in 1997 in order to align with those of the USA and EU.

8.1 Governmental Guidelines or Legislation?

In Japan, most research regulations established by governmental administrative guidelines are not legally binding. In addition, these guidelines were created by each governmental department to target various research procedures and medical care. Japan has a tendency to avoid legislature, primarily because establishing new legislation in Japan is incredibly difficult, and once established, it is then very dif-

© The Author(s) 2020
A. Akabayashi, *Bioethics Across the Globe*,
https://doi.org/10.1007/978-981-15-3572-7_8

ficult to change. On the other hand, administrative guidelines are flexible and can be changed, adapted, or adjusted. Because those who breach guidelines are penalized by strict sanctions such as loss of public funding, administrative guidelines are considered powerful within Japan. However, administrative guideline regulations have been applied unsystematically, and a new regulation is created for a given research field or medical procedure every time a new issue arises.

Consistency between guidelines is also problematic, leading to confusion onsite. The Japanese government structure is one of "vertical segmentation," in which the roles of each governmental agency are specified and problems are addressed with low rates of collaboration and slow methodology through interrelationships between multiple agencies. In such a system, guidelines are often produced independently by different ministries (e.g., the Ministry of Health, Labour and Welfare (MHLW), the Ministry of Education, Culture, Sports, Science and Technology (MEXT), and the Ministry of Trade, Economy and Industry (METI), to name a few) and relevant departments, leading to further confusion. Because of these problems, some recent guidelines have been issued by multiple ministries; this is a positive trend [1] (Table 8.1).

8.2 Ethics Committees in Japan

I have served as the Chair of the human research ethics committees at Kyoto University (4 years) and University of Tokyo (16 years). In what follows, I draw, in part from my own experience.

8.2.1 Number and Status of Ethics Committees

The first ethics committee in Japan was established in 1982 at Tokushima University Medical School.

The number of ethics committees at universities and hospitals conducting research in Japan increased dramatically as the twenty-first century began (Fig. 8.1) [2], due to the fact that administrative guidelines noted above required the establishment of an ethics committee within each institution.

The ethics review committee system was created on a voluntary basis at each institution during the initial phase (1980s to 1990s). Some were independent, approving research protocols. Some were involved in designing hospital policies for new technology. By the time the government administrative guidelines were established in 2000, research ethics committees are clearly designated as an

Table 8.1 Recent guidelines on scientific and ethical standards in Japan

Year	Administrative legislation	Affiliated ministry(s)[a]	Classification
1989	Guideline for good clinical practice (GCP)	MHW	Circular
1994	Guideline for clinical research on gene therapy	MHW	Notification
1996	Pharmaceutical affairs law (amended)		Law
1997	New guideline for good clinical practice (New GCP)	MHLW	Ministerial ordinance
1997	Organ transplantation law		Law
2000	Human cloning prohibition law		Law[b]
2001	Ethics guidelines for human genome/gene analysis research	MHLW, MEXT, METI	Notification
2001	Guideline for derivation and utilization of human embryonic stem cells	MEXT	Notification[b]
2001	Guideline for the handling of human embryos for research	MEXT	Notification[b]
2002	Ethical guideline for epidemiological research	MEXT, MHLW	Notification
2002	Guideline for clinical research on gene therapy (amended)	MEXT, MHLW	Notification
2002	Public health guidelines on infectious disease issues in xenotransplantation	MHLW	Notification
2003	Privacy protection law		Law
2003	Guideline for clinical research	MHLW	Notification
2004	Ethical guidelines for epidemiological research	MEXT, MHLW	Notification
2014	Guidelines on the derivation of human embryonic stem cells	MEXT, MHLW	Notification
2014	Guidelines on the distribution and utilization of human embryonic stem cells	MEXT	Notification
2014	Abolishing guidelines on clinical research using human stem cells	MHLW	Notification
2014	Ethical guidelines for medical and health research involving human subjects	MEXT, MHLW	Notification
2017	Clinical research act		Law

[a]*MHW* Ministry of Health and Welfare, presently the Ministry of Health, Labor, and Welfare; *MHLW* Ministry of Health, Labor, and Welfare; *MEXT* Ministry of Education, Culture, Sports, Science, and Technology; *METI* Ministry of Economy, Trade, and Industry

[b]English translations are available online: Human Cloning Prohibition Law (http://www.mext.go.jp/a_menu/shinkou/seimei/2001/hai3/4_houritu.pdf);
Guideline for Derivation and Utilization of Human Embryonic Stem Cells (http://www.mext.go.jp/a_menu/shinkou/seimei/2001/es/020101.pdf);
Guideline for the Handling of Human Embryos for Research (http://www.mext.go.jp/a_menu/shinkou/seimei/2001/hai3/31_shishin_e.pdf)

Fig. 8.1 The number of ethics committees established at general hospital in Japan

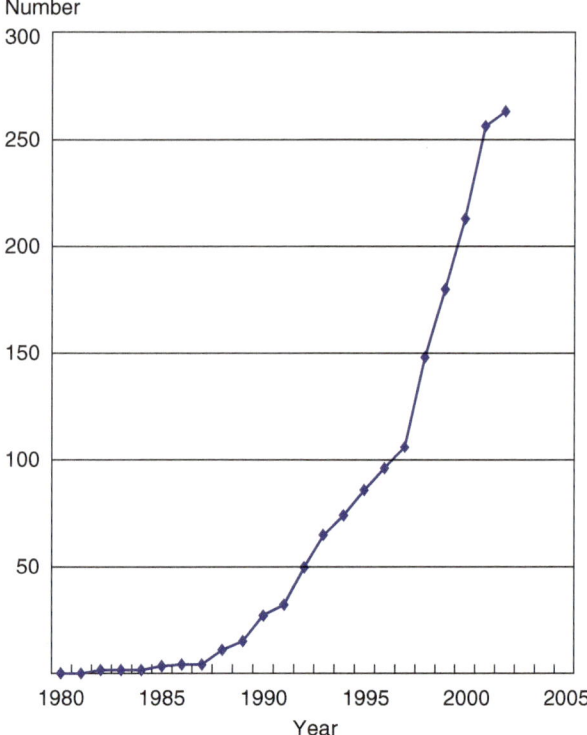

advisory body for the Dean or hospital director, and all research using human participants, or their material and data, are approved by research ethics committees.

8.2.2   Ethics Committee Members and Their Roles

Most research ethics committees operate voluntarily with regard to funding. Universities and institutions pay only small honoraria for external committee members. Internal members like myself have not received any recompense. I have chaired ethics committees for 20 years in a voluntary capacity. During that time I was on-call (24/7, 365 days a year) to confirm that informed consent of potential recipients of brain-dead liver transplants at Kyoto University Faculty of Medicine had in fact been given (See Chap. 2). In some instances I was called at 1:38 AM and had to be at the hospital by 5:00 AM (Fig. 8.2). Transportation costs were not covered by Kyoto University Hospital.

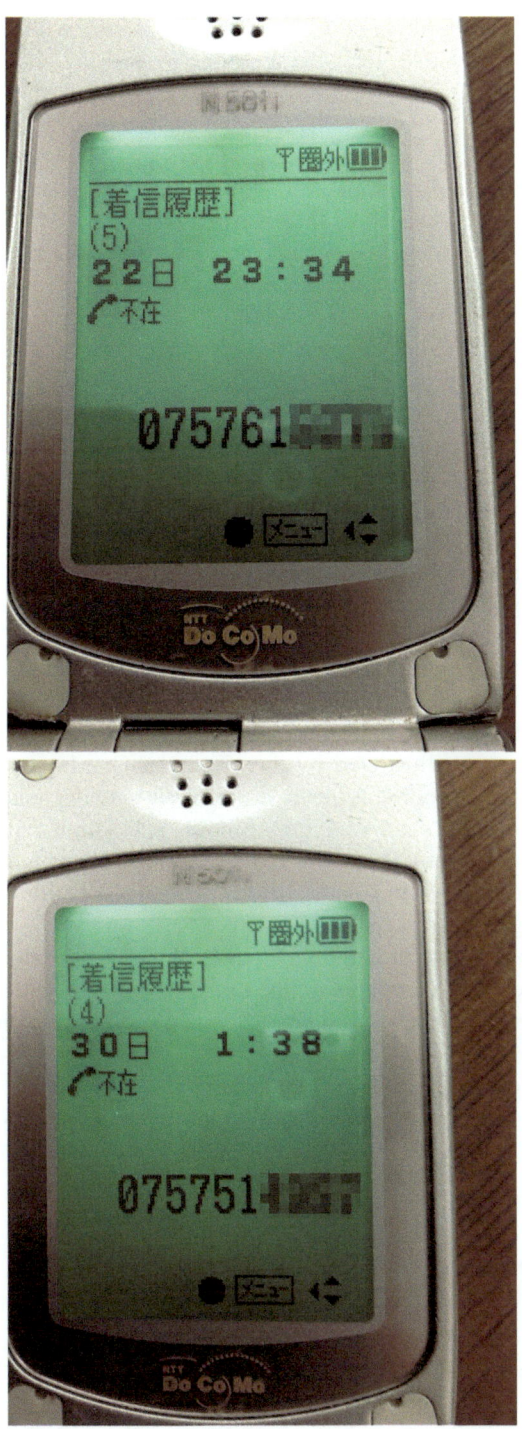

Fig. 8.2 Times the Chair of the ethics committee was called to confirm final informed consent recipient candidates from brain-dead donors

In order to create an ethics committee in accordance with the governmental guidelines, every committee needs to seek one or two lay committee members. We must ask, therefore, what is the role of the lay member? The ethics review system in the USA developed against a backdrop of racial discrimination and an inadequate level of patient advocacy. Therefore, ethics review committees required participation by persons who represented the perspective of the general public.

Japan imported this system and began using it without understanding this history or cultural context. Japan is a nation that is relatively racially homogeneous and has not faced a history of slavery or segregation. In addition, in the established system (Clinical Research Act, 2017) [3], patient rights were to be represented by a lawyer or someone with an understanding of bioethics.

My search for lay committee members has resulted in a professor emeritus several years into retirement, a retired employee of a pharmaceutical company, a principal of elementary or junior high school, a Buddhist monk, and a homemaker.

Within the flexible limitations of the administrative guidelines, each institution exercised discretion in determining their own suitable layperson committee member. This caused other problems, as described in the next section.

When recruiting lay committee members, many candidates asked me, "What am I supposed to do?" In response I suggested that they "Put yourself in the shoes of a patient or research participant, and when the explanation form or consent form is difficult to understand, or when you might, as a patient or participant, have a gut feeling that something is wrong, just speak up and say so."

The complexities are illustrated by this example: One lay committee member claimed that an informed consent form was difficult to understand. I listened carefully and suggested many changes. After spending some time reviewing it, she still claimed that it was unclear. I then asked her, "What changes do you think the researchers need to make to this part you mentioned?" She still had no answer and was unable to suggest an adequate alternative.

This phenomenon is not limited to lay members of ethics committees. For example, committee members from outside the university with backgrounds in biology ask questions in accordance with their own scientific interest, or simply display their knowledge.

In sum, many committee members did not understand what they were to discuss. It did not help that very few opportunities were available in Japan where these members could receive education about their role as ethics committee members.

8.3 Enforcement of the Clinical Research Act

The most recently established regulation is the Clinical Research Act (CRA), enacted in 2017 as cited above [3]. This law was created primarily in response to an incident of research misconduct that came to light in a collaborative research project between a university and corporation, in a manner similar to that of the Diovan incident [4].

This Act targets *only* specified clinical trials, which are:

1. clinical trials for pharmaceutical products that are either not yet approved or not approved for the particular purpose by the Pharmaceutical and Medical Devices Law
2. clinical trials for pharmaceutical products from a pharmaceutical company conducted with funding from the pharmaceutical company

The government explained that the reason the law applied only to specified clinical trials is that "excessive regulation can result in a weakening of freedom of research, so only a portion is subject to regulation." However, of the many types of studies being conducted, the law was applied only to two kinds of clinical trials, and the government required the independent establishment of a law-based institutional review board (*Nintei Rinshō Kenkyū Shinsa Iinkai*; hereafter, Certified Review Board: CRB) separate from the research ethics committees. This significantly increased the number of forms needed to report back to the government.

For the first time in Japan, members of the CRB were appointed according to legal statute. The CRB was to comprise:

1. A specialist from medicine or medical care
2. Either a lawyer or bioethicist (with an understanding of respect for human rights)
3. A layperson

Bioethicists and lawyers were pigeonholed together and by including the requirement of an "understanding of respect for human rights," the government intended that these individuals would also serve as patient advocates. Furthermore, there was an ongoing confusion about what the government's understanding of "a layperson." At the time, we (the administrative office of the CRB) sought a layperson committee member in order to create an institutional review board in accordance with the law, but upon sending the resumes of candidates to the MHLW, they would inevitably deem some of the candidates "inappropriate."

My search for layperson committee members yielded individuals as diverse as a professor emeritus, a retired employee of a pharmaceutical company, a principal of elementary or junior high school, a Buddhist monk, and a homemaker.

Some members of the MHLW argued that the professor emeritus had a conflict of interest with this particular organization (personal communication). They also suggested that a person who had retired from working at a pharmaceutical company could be considered to be medical personnel (personal communication). Their opinions on these matters seemed to vary each time our committee's officer inquired. Some argued that even a homemaker could be considered a specialist in the field of bioethics, if this particular individual were a member of the Japan Association for Bioethics. Thankfully, the Buddhist monk was never deemed "inappropriate."

Nonetheless, it might be argued that the Buddhist monk was in fact the most questionable person to act as a lay member Within Buddhism, there are various sects, each of which may have different stances on medical research. If a Buddhist monk is acceptable, then what about a Catholic priest? If a research proposal for embryonic stem cell is presented, would a Catholic priest not object to this? What would this Catholic priest say about research studies that use aborted fetuses?

In a multiethnic nation such as the USA, a history fraught with racial discrimination and strong religious opposition gave rise to the need for the ethics review system, including the inclusion of lay members. Importing this system with the simple objective to meet international standards (or, perhaps Western standards) creates operational problems, particularly on-site.

Furthermore, the format of ethics committee member composition of Japan's CRB is similar to that of Western nations. However, if an empirical study were conducted to compare the content being reviewed by these boards, some surprising conclusions might emerge.

8.4 Scientific Misconduct in Research: Cultural Perspectives on Criteria for Authorship

A researcher's competence in the field of biology has come to be judged by criteria including the number of publications, the impact factors (IFs) of the journals in which he or she publishes, and the number of times their papers are cited by others. The December 2018 version of the "Recommendations for the Conduct, Reporting, Editing, and Publication of Scholarly Work in Medical Journals (by the International Committee of Medical Journal Editors: ICMJE)" clearly defines the criteria that one should fulfill to be considered an author of a paper.

This issue is complicated by cultural differences [5]. Fetters and Elwyn compared the numbers of authors per original article by Japanese and non-Japanese research groups in two qualitatively similar medical journals (*Circulation Research* [IF 11.6, 2015] and *Japanese Circulation Research* [IF 4.1, 2015]) [6]. In each of 3 years, they noted that 2–3 more Japanese authors were included per original article published in *Japanese Circulation Research* than in *Circulation Research*. They attributed this difference to cultural differences in crediting authorship, highlighting the Japanese group-ethics, the role of professors in conducting research, and the funding system. They concluded that "the movement to credit only those who deserve authorship is noble, though the assessment of legitimate authorship is a cultural, not a scientific judgment [6]"

The Japanese group-ethic is certainly one cultural perspective. Although ICMJE stipulate the definition of authorship, cultural norms also play a part.

8.5 Conflict of Interest in a Society Supported by Fiduciary Relationships

Although the concept of conflict of interest (COI) overlaps between countries, the practice varies widely. Japan's governmental "Ethical Guidelines for Medical and Health Research Involving Human Subjects" (MEXT, MHWL) requires researchers to (1) ensure transparency, (2) include COIs in the protocol, and (3) explain COIs to participants. The Question and Answer section of these guidelines is also fairly simple. Only the Clinical Research Act (CRA) has a detailed discussion about COIs. In other words, research studies not under the CRA are monitored differently, usually according to institutional discretion.

Perceptions of COIs may differ, for example, between the USA and Japan, as the USA comprises a society rooted in contractual agreements, whereas Japanese society is based upon by fiduciary relationships. Social perception of COIs may also reflect whether a society is biased toward fiduciary relationships or contractual agreements.

An example of this is material transfer agreements (MTAs). MTAs are mandatory for all international collaborative studies conducted today. While Japan is not exempt from these agreements, I suspect that few deans in Japan actually read MTAs before signing them. Very few domestic studies in Japan *formally* require them. They do not need MTAs so long as the parties are in a fiduciary relationships. Written documents (MTAs) therefore do not have priority.

These apparent differences give rise to the question about whether or not COI guidelines should be tailored to the culture, society, and healthcare system in which they take place. Should every institutions policy stipulate "the percentage of option stock or amount of money received from an industrial sponsor," or rather demand that researchers "stay within a range accepted by social norms?" Japan should develop a set of guidelines considered internationally acceptable while still being suitable for Japanese society and culture. In this way, the systems in place used to monitor COIs represents an excellent model for comparing cultural and social structures [7].

8.6   An Addendum: Hospital Ethics Committee and Clinical Ethics Consultation

8.6.1   Clinical Ethics Consultation

In addition to institutional review boards (research ethics committees), Japan also imported the Hospital Ethics Committee system and clinical ethics consultation system. A similar format to that used in the USA was established, while embodying different values in the Japanese context.

A study comparing the USA and Japan's Clinical Ethics Consultation [8], employing the exact same setting of a decision-making proxy for a patient with Alzheimer's disease, found that the content of advice differed by country. Differences were identified in recommendation and assessment between the American and Japanese participants. In selecting a surrogate, the American participants chose to contact the grandson (legally the most clearly designated person) before designating the daughter-in-law as the surrogate decision-maker. They made an effort to discern the patient's preferences and thereby obtain a suitable surrogate. In contrast, the Japanese experts assumed that the daughter-in-law (a more distant family member, but one who lived nearby) was the surrogate and asked her opinions on the matter, with the aim to obtain a best interest judgment.

Case (from Nagao et al. [8])

Name and age	Mrs. Mineko Sakata. Age = 92 SEX: Female
Diagnosis	Late-onset Alzheimer's disease.
Chief complaints	disturbance of consciousness, cognitive impairment, and dysphagia

The patient began to exhibit impairment of memory and orientation in 1998 and has been progressing ever since. Her family first brought her to our hospital in July 2000 when they discovered she was wandering about aimlessly and screaming in the middle of the night (ambulatory automatism). She was diagnosed with late-onset Alzheimer's disease.... The patient's family admitted her to X Elderly Care Facility in October 2005. While the patient was at the X Facility, she remained drowsy throughout the day and night..... The patient could not chew or swallow, which made oral feeding difficult. In July 2006, the patient was transferred to our hospital. Our staff has tried to tube feed her via a nasal gastric tube; but she persistently removes the tube. The patient is currently physically stable and is not considered to be at the end-of-life stage. As a result, we recommended a gastrostomy (percutaneous endoscopic gastrostomy) for total enteral nutrition.

Mrs. Fujiko Sakata, the patient's daughter-in-law, has expressed that she would not want any other medical treatments if the patient were unable to eat. The patient also has a grandson whose name is Joji, Mrs. Fujiko Sakata's son. His opinion is that a gastrostomy would be allowed if it can prevent his grandmother from dying of starvation. Joji is currently in the USA since he has worked there for a long period of time.

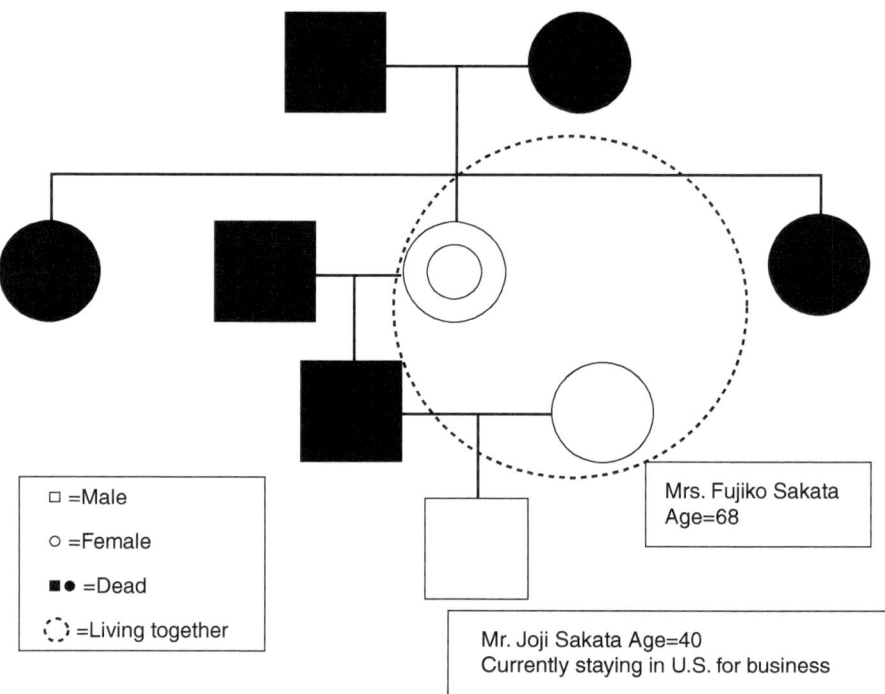

□ =Male

○ =Female

■● =Dead

⟨ ⟩ =Living together

Mrs. Fujiko Sakata Age=68

Mr. Joji Sakata Age=40
Currently staying in U.S. for business

8.6.2   University of Tokyo Model: Patient Relations and Clinical Ethics Center (PRCEC)

The structure of clinical ethics consultation (CEC) varies. After having observed many forms of CECs in other countries, I developed a unique CEC model for the University of Tokyo Hospital (1200 beds) [9].

The most prominent characteristic of my model is the combination of the patient complaint window with the CEC window. When a case is brought to the Center by a patient or hospital staff member, a nurse with extensive clinical experience and training in medical ethics serves as a gatekeeper and directs the case either to the complaints team or the ethics consultation team. The complaints team is made up of two nurses and three administrative staff members. If the case is assigned to the ethics consultation team, it is initially handled by the Center's vice director, a physician. He or she collects information and responses when the case requires an urgent response or, conversely, is relatively simple. Cases involving complex problems are handled by a CEC team composed of nurses, ethicists, and legal scholars from outside the medical school. Thus, the Center decides on a case-by-case basis whether to use an individual consultant or a team for ethics consultations. More complicated cases and those related to hospital policy are brought to the formal Hospital Ethics Committee.

So far, the Center handles approximately 2500 cases annually, among them about 5.0% were CECs.

The integration of CEC to handle patient complaints has relevant implications. First, this model ensures that patients and family members have free access to CEC services. At the PRCEC, about 25% of CEC services are used by patients or their families, indicating that this model is efficiently able to identify patient concerns (Fig. 8.3). Patients need not to determine the appropriate office to visit when they need advice or support. Furthermore, patients and family members must be able to, on their own, understand which problems are "ethical." Many patients and family members may not know what defines an ethical issue. This may be one reason why, although CEC services are available to patients and their families in the USA, use of these services remains low. The University of Tokyo model provides easy patient access to the relevant services, which reduces the burden of selecting the appropriate window.

The second implication of integrating CEC services with patient complaints is that it makes it easy to identify "ethical" issues within patient complaints brought to the office. A study conducted in the USA found that patient complaints covered a broad range of issues including communication problems, conflicts between patients and medical practitioners over treatment and care, and issues related to rights, such as confidentiality and informed consent. Some of these may be ethical issues, and in some cases, it may be more appropriate to conduct a CEC rather than treat them as

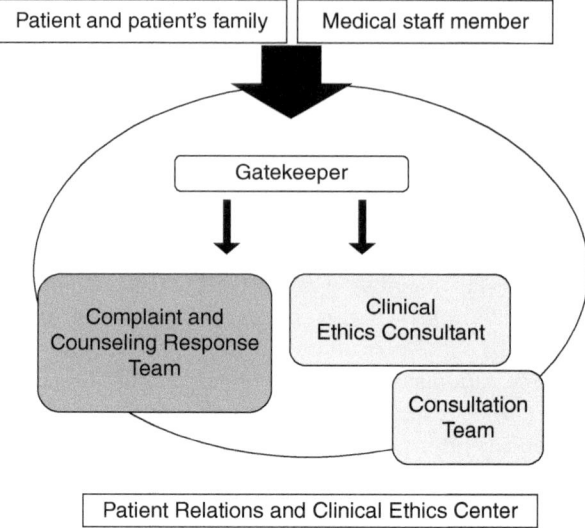

Fig. 8.3 Organization flow of PRCEC

mere complaints. In the USA, the Patient Advocacy Office is distinct from the Hospital Ethics Committee (HEC) or other offices that provide CEC services. Similarly, the Patient Advice and Liaison Service in the United Kingdom is separate from the division that provides CEC services. The PRCEC handles at least as many cases as USA hospitals of the same size, which provides some evidence of the efficacy of this model to identify ethical problems. The University of Tokyo model was presented at an international CEC conference and was well received. I hope that others involved in CEC systems would seek to develop more effective and appropriate frameworks that are in line with regional and institutional settings.

References

1. Slingsby BT, Nagao N, Akabayashi A. Administrative legislation in Japan: guidelines on scientific and ethical standards. Cambridge Q Healthcare Ethics. 2004;13:245–53.
2. Akabayashi A, Slingsby BT, Nagao N, Kai I, Sato H. An eight-year follow-up national study of medical school and general hospital ethics committees in Japan. BMC Med Ethics. 2007;8:8.
3. Akabayashi A, Nakazawa E, Akabayashi A. Implementation of Japan's first clinical research regulatory law: background, overview, and challenges. HEC Forum. 2019;31(4):283–94. https://doi.org/10.1007/s10730-019-09379-3.
4. Lancet Editors. Retraction--Valsartan in a Japanese population with hypertension and other cardiovascular disease (Jikei Heart Study): a randomised, open-label, blinded endpoint morbidity-mortality study. Lancet. 2013;382(9895):843. https://doi.org/10.1016/S0140-6736(13)61847-4.
5. Akabayashi A. Misrepresenting research. In: Thomasma DC, Kushner T, editors. Ward ethics. New York: Cambridge University Press; 2001. p. 249–51.

6. Fetters MD, Elwyn TS. Authorship. Assessment of authorship depends on culture. BMJ. 1997;315(7110):747.
7. Akabayashi A, Slingsby BT, Takimoto Y. Conflict of interests: a Japanese perspective. Cambridge Q Healthcare Ethics. 2005;14:277–80.
8. Nagao N, et al. Clinical ethics consultation: examining how American and Japanese experts analyze an Alzheimer's case. BMC Med Ethics. 2008;9:2.
9. Takimoto Y, Akabayashi A. Clinical ethics consultation in Japan: The University of Tokyo model. Asian Bioeth Rev. 2011;3(3):283–92.

Chapter 9
Modern Medical Professionalism

Abstract In this chapter, I will examine whether or not medical professionalism should take the same form worldwide. Japan has its own culture and ethos, both of which have significance in the clinical setting. However, if a Japanese doctor graduated from a Japanese medical school which is not accredited by international (Western) standards, then the doctor will not be able to work in the USA after 2023. Japanese medical schools are concerned about international standards because "Professionalism" is one of key part of accreditation. However, the question remains: should medical professionalism be measured in a universal and internationally standardized way? How would, for example, Japanese medical schools teach their medical students about the concept of autonomy, for which so many interpretations are possible? (see Chap. 3).

Further, I will explain the difficulties of teaching medical professionalism to medical students and young residents. In this chapter, I present two actual cases to illustrate these points. I bring to this discussion a case of medical professionalism using the Fukushima nuclear power plant accident as an example. Specifically, I question whether physicians are obliged to stay in an area highly contaminated with nuclear radiation. I also discuss whether one's obligation as a physician might require them to provide care during disasters. This is a different type of discussion about medical professionalism from those focused around clinical ethics.

Finally, I examine the happiness of the healthcare professional, a subject that has received little attention in the literature thus far in any country. I argue that the happiness of the healthcare professional should also be an important part of medical professionalism.

9.1 The Diversity of Medical Professionalism

How does culture affect the diversity of medical professionalism, and how much of this diversity should be acknowledged or accepted?

© The Author(s) 2020
A. Akabayashi, *Bioethics Across the Globe*,
https://doi.org/10.1007/978-981-15-3572-7_9

In 2002, several authors from Western nations published the Physician Charter. This Charter presented three fundamental principles (the principle of primacy of patient welfare, the principle of patient autonomy, and the principle of social justice) and ten professional responsibilities.

In September, 2010, the Educational Commission for Foreign Medical Graduates (ECFMG) in the USA announced that eligibility requirements for the examination to receive a physician's license in the USA restrict the pool to those who graduated from medical schools that are accredited to international standards, and that this accreditation should be mandatory after 2023. That said, in order to be accredited internationally, Japan must meet the standards of medical education set by Western nations. This created pressure, as revealed by the following statement issued in 2014 by the Ethics/Professionalism Committee Chair of the Japan Society for Medical Education: "At this point, medical schools nationwide are increasingly active in their movement toward international accreditation. In order to be accredited, they must fulfill the standards of medical education in Western nations; 'Professionalism' is one of the requisite items for medical education outcomes in the West."

The Physician Charter was translated into Japanese soon after publication, as were the Casebook on Human Dignity and Human Rights (UNESCO, Casebook Series, No. 1, 2011) and the Casebook on Benefit and Harm (UNESCO, Casebook Series, No. 2, 2011).

The "Clinical Ethics Education Package" (2016) published by the Japan Society for Medical Education has a thorough list of educational tools that looked remarkably Westernized. The core curriculum by the Ministry of Education, Culture, Sports, Science and Technology (revised in 2016) is shown in the footnotes. This curriculum is also "globally standardized."[1]

---

[1] Ministry of Education, Culture, Sports, Science and Technology
Medical Education Division, Higher Education Bureau
Core curriculum for medical education
Revised in 2016
A. Basic credentials/skills required of a physician
A-1. Professionalism
The physician must be thoroughly aware of their path of duty as a physician to protect others' health and be deeply involved in preserving human life. While practicing patient-centric medical care, they must work to master their role as a physician.
A-1-1. Medical ethics and bioethics.
Aims: To learn the importance of ethics in medical care and medical research.
Learning objectives: (1) To be able to present an overview of the historical flow of medicine and medical care and understand its meaning. (2) To be able to present an overview of ethical issues associated with clinical ethics, as well as issues related to life and death. (3) To be able to present an overview of the ethical rules established, including the Hippocratic Oath, the Declaration of Geneva, Physician's Occupational Ethical Guidelines, and the Physician Charter.
A-1-2. Patient-centric perspective.
Aims: To protect the secrets of the patient and their family, and while fulfilling one's duty as a physician and upholding medical ethics, prioritize patient safety above all else, constantly taking a patient-centric stance.
Learning objectives: (1) To be able to explain the basic rights of the patient as set forth in the Declaration of Lisbon on the Rights of the Patient. (2) To be able to explain the significance of the

Meanwhile, the Ottawa Conference (2010) emphasized the need to consider cultural context [1]. The general principles developed at this conference stipulate that "Professionalism is a concept that varies across historical time periods and cultural contexts." The sixth principle states, "Professionalism, and the literature supporting it to date, has arisen predominantly from Anglo-Saxon countries. Caution should be used when transferring ideas to other contexts and cultures. Where assessment tools are to be used in new contexts, re-validation with attention to cultural relevance is imperative." (p. 356)

The authors of the Physician Charter are all from the USA and Europe, so naturally, they depend on Western paradigms. In order to align Japan's standards with international standards, how should the Principle of Patient Autonomy, for example, be taught to medical students in Japan? As I explored in Chap. 3, the concept of autonomy in Japan differs markedly from that of the West, so is it appropriate simply to translate this and use it for education? Young medical students may interpret the principle of patient autonomy to mean that the physician should do whatever the patient asks within the bounds of clinical acceptability. However, is this the physician figure that is truly required in Japan's culture of *omakase*? Therefore, I wonder how much of this diversity should be acknowledged or accepted in medical professionalism.

In Japan, medical humanities education uses films and video educational resources. Those from English-speaking nations include "DAX's Case," "Discussions in Bioethics," "Awakenings," and "Patch Adams."

There are many different film educational resources in Japanese which take a different perspective, several of which are described below:

- *Ikiru* (To Live): (Director, Akira Kurosawa). This film was awarded the Special Prize of the Berlin Senate at the 4th Berlin International Film Festival in 1954. *Ikiru* grapples with the theme of receiving a cancer diagnosis. While the current discussion surrounding diagnosis disclosure is nearer resolution, at the time the film was presented, a diagnosis of cancer was not disclosed to patients. Its modern-day implications remain, however, as the viewer is able to view how the main character lives the remainder of his life. As such, it remains a highly valuable educational film.

---

patient's right to self-determination. (3) To be able to understand the patient's values and advise them accordingly, even in instances where various options are available, supporting the patient's self-determination. (4) To be able to explain the significance and necessity of informed consent and informed assent.

A-1-3. Duty and discretion as a physician.

Aims: To act with compassion and a deep awareness of the dignity of life and be aware of one's duty as a physician to protect human life and health.

Learning objectives: (1) To become capable of constructing trusting relationships with patients and their families in the clinical practicum for participatory medical examination. (2) Recognize that the values and social backgrounds of patients and their families can be diverse and be capable of responding flexibly to any and all of these. (3) To be able to explain why physicians must recommend the most suitable medical care for the patient. (4) To be able to explain that physicians are limited in their diagnosis and treatment depending on their own skills and the environment. (5) To be able to recount one's legal obligations as a physician and demonstrate these in practice.

- *Akahige* (Red Beard): Directed by Akira Kurosawa. This film was awarded the San Giorgio Award at the 26th Venice International Film Festival in 1965. *Akahige* remarkably reflects the saying, "Medicine is a benevolent art." This phrase reflects a long-held concept in medical ethics and is consistent with the Hippocratic Oath. In other words, it centers on paternalism.[2]
- *Okuribito* (Departures): Directed by Yōjirō Takita, this 2008 film was awarded the Best Foreign Language Film at the 81st Academy Awards. The film features an encoffiner as the main character and portrays a very uniquely Japanese perspective on corpses.

Overall, "To Live," "Red Beard," and "Departures" are all very moving films. Why did these movies win international awards? They are somewhat paternalistic and exotic for foreigners, and very Japanese. How did they impress a foreign audience?

Once again, Japan uses its unique path to obtain "international accreditation." I feel that this is acceptable, because the primacy of patient welfare remains intact. In educational settings, I teach that the Hippocratic Oath is still important. This is because the term "paternal" includes the concept of "benevolence," which remains an important value. If medical caregivers lack a sense of being "for the patients" in their professionalism, then what patient will come to receive medical treatment? In addition, regardless of the national culture, if medical caregivers lose this major concepts of practice, then where are they to place their occupational identity?

Overall, as long as the medical professional curriculum includes the principle of primacy of patient welfare, the principle of patient autonomy, and the principle of social justice, then sufficient diversity might be preserved.

9.2 Difficulties in Teaching Medical Professionalism to Young Students and Residents

At 20 years or older, medical students are all adults, and very few are likely to change their moral sensitivity or moral reasoning as a result of their medical education [3].

---

[2] "Medicine is the art of human-heartedness" is a phrase defined by the *Kōjien* (the mainstream Japanese dictionary) as "medicine is a benevolent/philanthropic road to save human life." It was used quite frequently, particularly during the Edo period, but the philosophical foundations are said to date back as far as the Heian period. Kaibara Ekiken (1630–1714), a Confucian scholar during the Edo Period (1603–1868), wrote the following in his "*Yojokun*" based on Confucianism [2]:

Medicine is the art of human-heartedness. A physician should build the foundation of his practice on human-heartedness and love, both of which focus on helping others. His intentions should not focus on his own profit and welfare. As this is an art of aiding people—who have been given their birth and nourished by Heaven and Earth—and takes charge of their life and death, you could say that a doctor is one of "humanity's officials." This is an extremely important position.

Teaching professionalism to medical students and young physicians is very difficult. Let me share an experience from 1985 when I was a Chief Resident at a general hospital with 400 beds in a rural area and was in charge of two first-year residents and one second-year resident [4].

Scenario One

A first-year resident A, who was in charge of a terminal female cancer patient with hepatoma in her 60s, was facing his first experience of a patient's death as an attendant physician. Her breathing became weaker and weaker and her blood pressure started to drop... A couple of days before, I had instructed him about how to tell when she had passed....by checking her corneal reflex, heartbeat, and respiration, and then to give the precise time of death to the patient's family. On that day, her husband, children, and relatives were with her, waiting for her last moments. The patient ECG monitor became flat. Suddenly, the resident started cardiac massage and asked a nurse to prepare adrenalin for intracardiac injection. I was a bit upset and after a few seconds, whispered to him not to do any more. In response, the resident answered "But Dr. Akabayashi, it is a physician's duty to do everything possible! I must do this." He performed the injection and continued cardiac massage despite my instructions to the contrary. After several minutes of attempted resuscitation, one of the family members firmly requested, "Doctor, please stop." Then I held the resident's arm and made him stop the massage.

Scenario Two

A male patient in his 70s with terminal pulmonary emphysema. He had been unconscious and on a respirator with a tracheostomy for more than 2 months. The attending physician was a second-year resident B. Because of malnutrition, hypoalbuminemia, and longtime bed rest, his face was awkwardly edematous, and according to his family, he looked like a totally different person. The patient had been aggressively treated every time he developed respiratory or urinary infections....Since B was in his second year, and was a competent practitioner, he did not need the detail of the treatment regimen to be intimately supervised. (Here I mean, the choice of drugs, content of infusion, and the setting of respirator.) One day, I suggested to him that he should talk with the patient's family, and discuss and reassess the patient's treatment plan. I also expressed my opinion that the treatment should be less aggressive. In response, B replied, "But Dr. Akabayashi, the patient does not have any malignant disease!" The treatment was continued as before, and the patient died about 3 months later.

There was no particular obligation for me to intervene with the resident's decision in either case. Their acts were not illegal. All I could say was that they prolonged life in an inappropriate way. Today, 30 years later, we no longer encounter these situations.

When should one intervene with a doctor in training's decision-making? I can think of only two situations when intervention by a physician in training is permissible, or even required. The first is when an action of a colleague is clearly violating the law. The second is when an act is against established hospital policy.

Medical training remains a type of apprenticeship. During the training period, young physicians need to learn the technical skills necessary to treat patients, but this is only part of their medical training. They also need to learn the skills that will enable them to resolve the complicated problems of medical practice. Those skills, I would argue, stem from an education in medical professionalism.

The value most sought after by Japanese people after World War II was longevity. Today, Japan leads the world in terms of mean life expectancy. From the 1970s through the early 1990s, physicians assumed a stance that prioritized longevity over QOL. Palliative care had yet to be developed. Medical professionalism at the time comprised, at the very least, primacy of patient welfare as interpreted by the physician, in other words, paternalism. Medical professionalism would change according to the change of the goals of medicine. By 2050, other forms of medical professionalism will have developed. Thus, I conclude that medical professionalism must differ by era and region. The most important objective of medical professionalism is for medical personnel to be always concerned about what is best for the patient (specifically, to align to the goals of medicine in temporal and regional context).

9.3 Emerging Issues in Medical Professionalism

Let us consider an emerging problem in medical professionalism. In what follows I consider patient welfare, patient autonomy, and social justice. However, it is "professional responsibility" that may be the most applicable principle in this discussion.

In the *Cambridge Quarterly of Healthcare Ethics*, the section editors for Professionalism discussed my articles in their Introduction: A Modern Version of an Ancient Question, as follows [5].

> With remarkable candor, Dr. Akira Akabayashi acknowledges that he responded to a colleague's request for his ethical opinion on this question from the relative safety of Tokyo, when the question involved the professional commitment of a physician just outside the official 20-kilometer evacuation zone surrounding the damaged nuclear reactors at Fukushima. His commentary has a vividness that approaches the drama of real-time deliberations and ends with a note of uncertainty…

> Nevertheless, their survey is helpful in placing Dr. Akabayashi's article in a historical context that extends far earlier than the outbreak of SARS, to which Dr. Akabayashi turns in a search for reflection and precedent. We would like to extend to our readers an invitation to respond in future issues of the *CQ* Professionalism section to the questions Dr. Akabayashi has posed, and that, as he suggests, continue to present challenges for bioethicists.

In my article, "Must I stay? The Obligations of Physicians in Proximity to the Fukushima Nuclear Power Plant," I question whether a physician is obliged to care for their patients if doing so would mean that they themselves are risking danger [6] (open access).

Case Dr. N's Request to Leave
In a hospital located a little more than 20 km (evacuate zone) from the affected nuclear reactors. I was asked the following question by the director of the hospital:

"a young female physician, Dr. N, wants to go home to Hiroshima, but we don't have enough physicians at the hospital. If she leaves, there will be no one to take care of her patients and the evacuees. What should we do?"

Dr. N argues that: "My parents are pleading with me to come back home to them in Hiroshima. I have a family that needs me. What are you going to do for me if I can't have any more children because of this? I can't continue treating patients under threat of contamination. I have the right to escape. Indeed, the Americans are evacuating, aren't they?"

Does Dr. N have a duty to remain, despite her reasons for leaving? Her superiors insisted that she stay, but does she have an obligation to comply? Should the hospital demand that she remain treating her patients?

9.4 On the Happiness of Medical Caregivers

I will conclude with a discussion that is often overlooked in publications on medical professionalism, and medical ethics. That is, what does happiness means to the physician (or medical caregiver)? Many articles discuss the obligations and responsibilities of physicians, but this issue is hardly ever addressed [7].

When teaching, I ask my students to consider that the core ethical question is, "what defines better medical care?" and "what comprises a better patient/medical caregiver relationship?" Indeed, medicine should be practiced in such a way that the medical caregiver is constantly thinking about the best possible benefit for the patient. However, in recent medical care settings, due to the collapse of the medical care system and the existence of "monster patients" (those who make serious complaints or who are verbally or otherwise abusive), it is also the case that medical caregivers who have been scarred by past incidents are in situations involving an unforeseeable future. The collapse of the medical care system in Japan has come at the end of rapid developments in modern medicine. Burdens, therefore, are currently shouldered by overworked medical caregivers, among whom depression is common and rates of suicide are increasing. This medical care setting does not represent an environment in which medical caregivers have the time and space to always work in their patients' best interest

The question therefore remains: What is needed for medical caregivers to be happy? A word of thanks issued from a patient is said to be one of the best things to make medical caregivers happy. In addition, the feeling that health workers and doctors are healing diseases, removing pain, and are contributing something to the patients and to society undoubtedly serves as support.

However, even if we begin with Aristotle's Nicomachean Ethics and eudemonism and continue with more contemporary work concerning happiness theory, in particular drawing on Alain [8], Russell [9], and Hilty [10], we must look at happiness in a wider context.

At this historical moment, at its basis, happiness among medical caregivers is borne from the relationship between patient and caregiver. However, in the future, this may change as new ethical issues emerge. However, as long as the human race continues medical care will exist, and thus ethics in medicine will be necessary. In addition, regardless of the era, even in the era of artificial intelligence-based medicine, medical personnel must "heal and save patients."

I would like to leave the reader with a last idea. Namely, proper consideration of "the happiness of the medical caregivers" will expand the breadth of medical professionalism, with broad implications for the future.

References

1. Hodges BD, et al. Assessment of professionalism: recommendations from the Ottawa 2010 Conference. Health Teacher. 2011;33:354–63.
2. Ekiken K. YOJOKUN: life lessons from a Samurai, translated by William Scott Wilson. Tokyo: Kodansha International; 2008. p. 197.
3. Akabayashi A, Slingsby BT, Kai I, Nishimura T, Yamagishi A. The development of a brief and objective method for evaluating moral sensitivity and reasoning in medical students. BMC Med Ethics. 2004;5:1.
4. Akabayashi A. To intervene? In: Thomasma DC, Kushner T, editors. Ward ethics. New York: Cambridge University Press. p. 244–7, 2001.
5. Wicclair M, Barnard D. Professionalism: introduction: a modern version of an ancient question. Camb Q Healthc Ethics. 2012;21(3):391.
6. Akabayashi A. Must I stay? --- the obligations of physicians in proximity to the Fukushima nuclear power plant. Camb Q Healthc Ethics. 2012;21(3):392–5.

7. Akabayashi A. The concept of happiness in oriental thought and its significance in clinical medicine. In: Japanese and Western bioethics, philosophy & medicine series 54. Netherlands: Kluwer Academic Publishers; 1997. p. 161–4.
8. Alain, Propos sur le Bonheur, Gallimard, Folio, 1995.
9. Russell B. The conquest of happiness. Allen and Urwin; 1930.
10. Hilty C. Gluck. Huber & Co.; 1981.

Chapter 10
What Does It Means to be Truly "Interdisciplinary"?

Abstract Before concluding this work, let us return to some bioethical theories. The theme of the present chapter is integral to bioethics. My main intent is for each reader to revisit his or her definition of the meaning of "interdisciplinary," a core term in discussions of bioethics. The manner in which this term is used varies widely. At the simplest level, "interdisciplinary" is used to indicate that researchers and others from multiple academic fields have collected together their own individual theories on a particular topic. However, it is worth wondering how much each researcher actually understands the writings and thoughts of those in other fields. In the present chapter, I first ask what is required to be truly "interdisciplinary" and present a sport ethics article my colleagues and I wrote as an experiment to demonstrate these points. My hope is that my readers will consider how this article could be changed in order for it to be understood better by as many readers as possible.

Bioethics is often said to be an interdisciplinary field of study. However, "interdisciplinary" is a complex term. In the initial stages of the debate on brain-death in Japan, it was quite typical for symposiums comprising researchers and others from multiple fields to begin with "from the standpoint of medicine," "from a legal perspective," or "from a philosophical point of view," before presenting their own opinion from the specialty field. However, as various opinions were voiced from different fields, this approach was considered "interdisciplinary." Unfortunately, this approach cultivates a very shallow level of debate. This, in turn, means that valid interactive conversations never begin. At the time of the brain-death debate, communication skills within science and technology had not yet developed in Japan, and there was little that could be done when facing this first major problem in bioethics.

A truly interdisciplinary conversation will never begin if we merely listen to the perspectives of the specialists, but then investigate the issue no further. So the question remains: what does it mean to be truly interdisciplinary? I feel that truly interdisciplinary dialogue implies a particular posture taken when addressing a given problem. Thus academic debate should result in the participants achieving a deep

© The Author(s) 2020
A. Akabayashi, *Bioethics Across the Globe*,
https://doi.org/10.1007/978-981-15-3572-7_10

understanding of each other's opinions, and even if a resolution is not achieved immediately, obtaining the sense that "the discussion moved forward/the understanding of the other person has deepened." Dialogue can only begin with a general understanding. As the dialogue begins and the discussion continues, a mutual understanding of each participant's views is deepened further, and the result is something that might be considered truly interdisciplinary. Nonetheless, it is difficult to define exactly that what is "truly interdisciplinary," but one prerequisite might depend on the "attitude" of those involved while conducting the dialogue.

In the present chapter, I will give an example. The paper below has not been published elsewhere. Using sumo wrestling as an example, one author wrote the first draft without limiting the argument to any one academic field. After the first draft was created, other authors from a variety of specialties added the flesh to the skeleton. All co-authors consented to the publication of this article in the present text. It is written with terminology from ethics, philosophy, sociology, law, psychology, and anthropology. Scholars in some specialties may criticize this as superficial. However, as a discussion increases in specialty, more specialized terminology is used such that some researchers may not be able to understand sufficiently the writings of their colleagues in other fields. One other criticism may be, "Well, that's somewhat interesting, but you need to deepen the discussion." However, to "deepen the discussion" in one specialty field would make this less interesting to those in other fields.

Sports ethics, which has a slightly different feel from the other themes mentioned thus far, is becoming an important field within bioethics. I hope that my readers will consider the relevance to the objectives of the text below as we discuss the topic of sumo wrestling, the national sport of Japan.

Original Article

Do Professional Athletes Have the Right to Dispute a Referee's Judgment? An Ethical Analysis of Sumo Wrestling in Japan

Akira Akabayashi, Akifumi Shimanouchi, Eisuke Nakazawa, and Aru Akabayashi
    Department of Biomedical Ethics, The University of Tokyo Faculty of Medicine, Tokyo 113-0033, Japan.

Abstract On November 22, 2017, the *yokozuna* grand champion Hakuho (a Mongolian citizen), losing during the final bout of the day, thought that the initial charge (*tachiai*) was incorrect and raised an objection with the referees. His objection was ignored by the referees, and Hakuho was subsequently subject to intense criticism for lack of dignity, rule violations, and foolish behavior. The following day, the judging department issued a severe warning to Hakuho, and he immediately apologized.

We first examine whether an athlete in modern sport has the right to dispute a referee's decision, in order to examine the concept of rights that are utilized herein, and discuss the characteristics of such rights. We then analyze how professional sumo is not a typical modern sport, and based on socio-ethical aspects, address the question of whether sumo should in fact join the category of modern sports. Finally, we argue that Hakuho's behavior after the incident can be justified under virtue ethics, and concluded that analysis of the Hakuho case could provide insight about the state and future direction of many world sports that occupy an uncertain space between traditional and modern sports.

Keywords Sumo, professional athletes' rights, modern sports, traditional sports, Japan

Introduction

The International Sumo Federation (ISF), in which 84 countries are registered, holds tournaments divided by weight class every year and also allows women to participate (http://www.ifs-sumo.org). The ISF is one of the International Sports Federations recognized by the International Olympic Committee (IOC). The ISF and the Japanese Olympics Committee have been working proactively to make sumo an Olympic sport [1, 2].

Sumo is divided into *Oh-sumo* (professional sumo) and amateur sumo. The ISF has jurisdiction over amateur sumo, while the Nihon Sumo Kyokai (NSK, http://www.sumo.or.jp/En/) exercises jurisdiction over *Oh-sumo*, which is the national sport of Japan.

While there are multiple theories about its origins, the history of sumo can be traced back to the eighth century. The sport has existed in various forms and contained elements of religious ritual. Modern sumo is said to have begun to converge during the Edo Period (1603–1868) [see, for example, 3–5].

At present, *Oh-sumo* has been internationalized to a significant degree. Of the 70 wrestlers ranked *Jūryō* or above at the March 2018 tournament, 18 (25.7%) were not Japanese citizens. Among the three *yokozuna* grand champions, two are Mongolian.

This paper takes up a recent case from *Oh-sumo* in order to discuss the rights of professional sumo wrestlers from an ethical viewpoint and offer perspectives on the future orientation of *Oh-sumo*.

The Yokozuna *Hakuho Case*

On November 22nd, 2017, the previously undefeated *yokozuna* grand champion Hakuho (a Mongolian citizen) lost to the *sekiwake* Yoshikaze (a Japanese citizen) in the final bout of the day. Thinking that the *tachiai* initial charge was incomplete, Hakuho let down his guard and was rammed out in one stroke by Yoshikaze. Dissatisfied, Hakuho raised his right hand to appeal to the referees, a gesture requesting review by referees called a *mono-ii*, and continued standing outside the ring (*dohyō*). He moved his right hand five times to appeal to the Shikihide referee (former *maegashira* wrestler Kitazakura), who was directly opposite of him. There was, however, no rematch granted. The referee urged him several times to ascend to the ring, and after 61 seconds he finally did so, only to again raise his right hand to appeal. Yoshikaze was declared the winner and stepped out of the ring. But for the next 17 seconds, Hakuho stood at full height in the ring, refusing to leave. After being urged several times to "step down," he finally left. The announcer for Nihon Hoso Kyokai (NHK) broadcasting the tournament commented, "This is the sort of thing that must never occur."(See Figs. 10.1 and 10.2, and the following YouTube video describing the course of events:https://www.youtube.com/watch?v=60DYeZgMJMU).

Hakuho, returning to the dressing room, was asked by the media "Did it seem like a false start (*matta*)?" and he replied, "Well, that's how it seemed. I wanted them to review it once (on the video). It isn't that I am unconvinced, but it is true that we were out of sync [6]."

Criticism of Hakuho began immediately after. The Shikihide stable master stated: "It is absurd as far as the rules go, isn't it? A *mono-ii* appeal can be raised by the referee or wrestlers waiting ringside. But this is inconceivable [7]." Hakkaku, Chairman of the NSK board, commented, "Wrestlers cannot make judgments on their own. It is unsportsmanlike [8]" In addition to criticism from the referees, Hakuho was widely criticized by the media in general, with the incident characterized as "the shameless behavior of a stubborn *yokozuna*" [9], and as "unthinkable behavior, for a wrestler to contest his own loss [10]." There were also countless critical posts on SNS and YouTube, including discriminatory statements such as "You see, this is the problem with Mongolians."

Fig. 10.1 Hakuho appealed to the referees and did not step up in the ring

Fig. 10.2 Hakuho, losing to Yoshikaze, appealed to the referees and did not step down from the ring

On November 23, the day after the match, Hakuho was called before the judging department and issued with a severe warning for his behavior following his loss to Yoshikaze, including appealing about an uncompleted *tachiai* and demanding a *mono-ii* discussion to review the referee's decision, which were called undignified behavior unbefitting a *yokozuna*[11]. In response, Hakuho took a repentant stance stating that he would "sincerely take it to heart and act properly in the future [12]."

Discussion

Does an Athlete in Modern Sports Have the Right to Dispute a Referee's Decision? If So, from What Standpoint Is That Right Justified?

Based on the classical framework of rights theory set forth by Hofeld, the "right" to make a "claim" has been understood as something interdependent with a "duty [13, 14]." In other words, if X has a right vis-a-vis Y, this means that Y has a duty to discharge vis-a-vis X. This argument typically posits a contractual relationship, which may be either written or social. With respect to the theme of our paper, the right for an athlete to dispute a referee's judgment in modern sports, however, we find it difficult to account for this right within that classical framework. We thus propose the concept of "**rights characteristic to modern sports**," and discuss the standpoints from which these could be justified.

To confirm the factual basis for the right of a competitor to appeal a decision, the Olympic Charter includes stipulations about appeals and the procedures for settling disputes, and the Court of Arbitration for Sports plays a central role [15]. FIFA has also set forth a Human Rights Policy [16]. What can be said based on these developments is that modern sports were founded on the basis of the concept of fundamental human rights that took shape during the eighteenth century.

Is the right to appeal the decision of a referee, then, counted among the rights within modern sports? If so, what sort of justification exists for these rights? In international tennis tournaments, for example, an athlete can challenge a referee's judgment. In numerous other sports, instant replay by video is permitted. In what follows, we discuss two standpoints, through which we hold that athletes in modern sports do in fact have the right to dispute the ruling of a referee.

The first standpoint is that of **accountability**. In general, when the reason for one's behavior is questioned by others, there is a responsibility to explain. In his philosophical analysis of referees, Collins describes that in sports, referees have two characteristics. First, they have ontological authority, and they can decide whether an athlete's action under a certain situation (for example, offsides in soccer) falls under the definition of foul play. Why do they have such authority? The answer lies in the second characteristic of referees: epistemological privilege. That is, a referee's judgment is regarded as a "superior view," since it is made from a suitable position (whether they stay in one place as in the case of tennis, or move around with the players as in the case of soccer). Also, referees are trained experts whose abilities are expected to improve through actual refereeing activities. Moreover, their qualifications and eligibility must stand up to the scrutiny of a group of specialists. Hence, it follows that the athlete should obey the referee's ruling, as it is based on "specialist skills" [17].

That said, "fallibility" is a general human trait, and not just limited to referees. It is because of the possibility of misjudgment that systems such as a video replay were introduced in modern sports. Inappropriate judgments by referees would make it difficult for modern sports to continue. Thus there must be mechanisms for objection and accountability. Athletes are obliged to obey the final ruling of a referee but they also have the right to appeal. On the other hand, while the referee has the authority to make a final decision, if the athletes raise an objection, he/she are obligated to confirm the propriety of the decision by appropriate means (e.g., a video replay system) and provide explanations to the athletes as well as to the audience (i.e., accountability).

It would be irrational for referees not to accept the athlete's appeal, if they are aware of their own fallibility in decision-making, as their authority would be lost if they were found to be prone to misjudgment. "Accountability" is another obligation they must fulfill if they strive to carry out their specialized job as referees appropriately. Accountability might even be described as a virtue of sorts, like fairness or modesty. In the end, modern sports in the proper sense cannot be founded on rules that do not incorporate an athlete's right to challenge a referee's decision, or referee accountability.

Let us consider more specifically referee accountability in the case of sumo. Current rules allow only ringside referees, and wrestlers waiting their turn to protest against or dispute (*mono-ii*) the ruling of the referee. However, from the perspectives of "epistemological privilege" and "superior view," the wrestlers who are participating in the match have witnessed the moment of winning/losing from a close distance, so they have equal (if not more) capabilities to the referee or ringside referees and wrestlers to properly judge their victory/defeat. In this regard, the current situation (i.e., the right to appeal to the referee in the form of *mono-ii* or a request to confirm on the video is not extended to the wrestlers in the match (as concerned parties)) might reflect that referee accountability is going unfulfilled.

The second standpoint is that of **fairness**, an essential value that makes possible modern sports. Modern sports might be called "a practice constituted by rules" [18, 19]. Whether it is market economics or sports, fair rules are essential for any kind of competition to exist as a practice. Fairness is both a value that must be practiced in modern sports, and an indispensable value that makes possible the very practice of modern sports. Furthermore, if fairness is internalized by athletes, this leads to the cultivation of a sportsmanship that values fairness while aspiring to individual excellence. The right of an athlete to appeal a referee's decision is thus also justified from the standpoint of fairness.

How does this affect the sumo case? Under the current rule, wrestlers are permitted to dispute a referee's decision regarding other wrestlers' matches, but not their own. As suggested above, in modern sports, referees' authority to make a final decision is paired with their accountability when their decision is challenged, just as an athlete's right to challenge a referee's decision is paired with their obligation to obey the final ruling. In this sense, it would be fairer to recognize the right of sumo wrestlers who participated in the match to raise a *mono-ii* appeal.

In what follows we will discuss what characterizes the concept of "**rights that are characteristic to modern sports**." What rights specifically are included among these, and from what standpoints are such rights justified? What should first be confirmed is that these rights come into being because modern sports find their basis in the concept of fundamental human rights. For example, in boxing, the athlete has the right to be protected by a referee from danger to life, from the standpoint of **nonmaleficence**, which stresses that one will not be subject to undue harm. The right to participate in competition regardless of race, sex, or religion is justified from the standpoints of **uniformity of opportunity** and **equality**. Furthermore, in recent years, athletes' rights of publicity (for example, in the case of female beach volleyball players and swimmers) have come to be given weight from the standpoint of **privacy**. In addition, the rights possessed by competitors are not limited to those that apply during competition. Athletes' rights not to be subject to improper treatment by instructors during practice (for example, sexual harassment or intimidation) may be recognized from the standpoint of **"respect for individuals."** Athletes participating in the Olympics who are dissatisfied with the propriety of the selection process, doping certification, or suspensions for rough play have the right to appeal to sports arbitration bodies. This is none other than the right to **"appeal."** This right to appeal has already been incorporated into the practice of modern sports, regardless of whether athletes actually exercise it. For this reason, disputing the decision of a referee, as part of an athlete's right to appeal in the broadest sense, is included within **"rights that are characteristic to modern sports"** from the standpoints of **accountability** and **fairness**.

Is Oh-sumo as Practiced in Japan Really a Modern Sport?

The sociologist Guttmann listed secularism, equality, specialization, rationalization, bureaucracy, quantification, and records as the seven characteristics of modern sports [20]. Thompson has discussed whether these characteristics can be found within modern sumo [21]. Thompson's analysis does not find that it fulfills all seven characteristics fully, although *Oh-sumo* has modernized to some degree. We concur with this conclusion.

Yet, with respect to Thompson's judgment that professional sumo has been to some degree **_Rationalized_**, we come to a somewhat different opinion when discussing the present Hakuho case. Rationalization refers to the process by which facilities and tools are standardized, and the rules are made universal and clearly stipulated in writing. Thompson has pointed to the standardization of the *dohyō* ring and the clear stipulation of rules. In actuality, however, the documentation of rules has not been sufficient.

In 1955, the NSK issued the "Official Sumo Rules." These were then revised 1958, but it is now impossible for ordinary people to obtain them. One of the authors visited the Sumo Museum (http://www.sumo.or.jp/EnSumoMuseum) and confirmed with the archivist that the rules have not been revised since 1958. Searching at the Japanese National Diet Library, we confirmed an entry including the "Official Sumo Rules" [22], and examined the contents.

After carefully examining the Official Sumo Rules, we found that there was no clearly documented stipulation that "competing wrestlers must not attempt a *mono-ii* appealing a referee's decision." All it did include in Regulation 5 on "Referee Regulations" were agreements about inspectors' *kensayaku* (=referees) *mono-ii* (Articles 4, 7, and 9) and a statement that wrestlers waiting ringside for the match could do a *mono-ii* (Article 5).

What should be noted here is that rules in general take the format of positive lists of matters that are permitted and negative lists of matters that are prohibited. Rules often contain a mix of these two formats. There is the view that because *mono-ii* is included on the positive list of the Official Sumo Rules, and foul play is noted on the negative list, it was not necessary to clearly stipulate that "competing wrestlers must not attempt a *mono-ii* appealing a referee's decision." Yet, the Shikihide stable master's comment that "It is absurd as far as the rules go, isn't it?" is not accurate [7].

There is no mention of *mono-ii* by competing wrestlers on either the positive or negative list of the rules. Therefore, it is that it is an unwritten rule. We assert then, that Thompson's understanding of sumo's rationalization is correct, but only to a degree. That is sumo has been rationalized when compared to the Edo Period. Thompson likely wanted to argue that sumo was rationalized because the Official Sumo Rules were created, but when we consider the Hakuho case in light of these rules, there is no sign that Hakuho violated any clearly documented rule.[1]

Guttmann argues that sumo is a "hybrid sport" as follows [23]:

> The result of these cross-current of modernization and what we might refer to as "tradition-alization" was the hybrid sport that we see today.......No matter. Sumo, like the imperial line that traces its origins back to the goodness of the sun, is authentically Japanese. No traditional sport—with the possible exception of Spanish bullfighting—has more success-fully "naturalized" its concessions to modernity.

This type of approach is in fact Japan's survival strategy. Not simply in sports but also in politics, scholarship, religion, and all social systems and products, whenever something is imported to Japan, it is modified and fused to be compatible with Japan, and "reconstructed" so that it can easily be adopted. Religious matters are an excellent example. With the introduction of Buddhism, there was a process that harmonized the new religion with native Shintō (the phenomenon of the syncretization of Shinto with Buddhism, which is distinct from polytheism). Professional sumo was transformed from a traditional sport into a hybrid sport, rather than a modern sport.

---

[1] On the subject of *Equality* in sumo, Thompson notes that, although women previously could not even watch sumo, they now can. In effect, he found trends of *Equality* in the modernization in *Oh-sumo*. Thompson carefully avoids judging whether sumo is modernized or not by using Guttmann's seven characteristics [21]. However, we have some concerns.

In *Oh-sumo*, the tradition holding that the *dohyō* ring is off limits to women persists. In 2000, during the March *Oh-sumo* tournament, Osaka's prefectural governor Fusae Ōta expressed her desire to present a Governor's Award during *Senshuraku* (last day of the tournament) by herself in the ring, but the NSK strongly disapproved. This became a widely publicized social issue, but the governor ultimately abandoned the plan.

The tradition still continues to this day. At just after 2 pm on April 4, 2018, during the *Oh-sumo* Spring Tour's "*Oh-sumo* Maizuru Tournament" held in Maizuru City, Kyoto Prefecture, Mayor Ryozo Tatami (67-year-old male) collapsed while giving a welcome speech. As several women were performing cardiac massage on the mayor in the ring, announcements were made at least three times saying "Women please exit the ring" and "Men please enter the ring [28]."

On the evening of April 4th, Hakkaku, the NSK chairman, admitted that the NSK's *gyoji* refer-ees made several announcements saying "Women, please leave the ring," and commented, "The *gyoji* made these calls because they were distressed, but it was not an appropriate response to a situation in which a human life was on the line. I deeply apologize [29]."

This comment by Hakkaku suggests the view that a human life overrides the value of tradition, which we agree with. However, NSK's position has not changed at all since the case of Ōta in 2000. On April 6, 2018, only two days after the Kyoto case, Mayor Tomoko Nakagawa (70-year-old female) of Takarazuka City, Hyogo Prefecture, was prohibited from giving a speech in the ring for the same reason as in the Ōta case in 2000. Mayor Nakagawa commented that "It is regretful I could not make my speech in the ring. While keeping the tradition, it is important to have courage to change [30]." This has become a social issue once again, but change seems unlikely. Does *Oh-sumo* reflect the form of society of this period as Thompson stated?

Thus, professional sumo has a hybrid dimension as described by Guttmann, but if we take into account characteristics such as the lack of universalizability, publicly open rules, and the inequality surrounding women's participation. In that sense, the tradition has been stubbornly preserved in a changed form, and there are, in fact many elements that have not been modernized.

Thus, our conclusion is that "professional sumo (*Oh-sumo*) in its current form is not a typical modern sport."

Should Oh-sumo Join Modern Sports? Professional Sumo and Cultural Imperialism

Cultural imperialism within modern sports should certainly be criticized [24, 25]. Sports have been employed in political contexts during the history of colonization. This undermines our understanding of sport as something good, as it promotes health.

On the culturally imperialistic dimensions of modern sports, Guttmann writes [26]:

> Standardized universality does replace diversity, but, when accompanied by the other characteristics of modern sports, it enables everyone to play the game—whatever game it is. .....As Ommo Grupe noted at the international symposium, modern sports are—despite their many abuses—inherently cosmopolitan.......If sports are an occasion for the expression of communitas, which they can be, let them express the human community as well as the tribal one.

This is a powerful ethical and normative claim. We, however, would like to express some concern with the way that Guttmann emphasizes the value of universality in modern sports.

Because we are not cultural anthropologists, we will not adopt a position of relativism. Nonetheless, we must not forget that traditional sports arose from games (amusement) and religious ceremonies. In the modern and contemporary period, robbing people of their freedom to play games, or their freedom of religious belief, would be a **violation of civil liberties**.

Guttman's claim should be limited to the context of "modern" sports only. In actuality, the "International Conference on Traditional Sports" was held in Tokyo in 1993, and declared for the first time how traditional sports could serve as a means of intercultural understanding on a global scale [27]. In the present day, as the world intensifies its internationalization even as the rise in nationalist sentiment emerges as a serious social issue, it is thought that traditional sports can play an important role in true international exchange and internationalism based in respect for other cultures.

For Japanese people, *Oh-sumo* is a popular national sport and form of mass entertainment. Who should decide, and on what basis, the question of whether *Oh-sumo* should join the club of modern sports. Thompson explains that "The form of sumo reflects the form of society in any given period. Since long ago *Oh-sumo* has adapted to society, and it is necessary for it to do so now as well [21]." In saying that sumo should reflect the direction of society in a particular age, he is making a powerful normative, ethical argument.

Our view, which is similar to Thompson's, is that because sumo has an important cultural dimension as a national sport, its rules and their application should take Japanese national opinion into account. Yet, currently one-fourth of ranking sumo wrestlers are foreigners, which suggests that the internationalization of *Oh-sumo* is already happening. Therefore, we would recommend that *Oh-sumo* should tackle with the issue of internationalization, in order not to remain a sport that is closed off from international society just because it is a national sport.

Furthermore, even though many, including the government, NSK, and citizens hope that sumo will be accepted as an Olympic sport in the future, this is not possible in its current form (sumo has already been rejected by the IOC multiple times). If sumo truly aspires to become an Olympic sport, it will likely be necessary to advance its transformation into a modern sport and, from the standpoint of fairness, to clearly document the rules and discuss the right of an athlete to appeal a referee's decision. At the same time, we would like to add that yet another key ethical consideration is that this process of decision-making should not be made solely by the NSK's board meeting behind closed doors, but rather in a way that adheres to **procedural justice** by taking into account a wide spectrum of opinions within Japan and abroad.

Can Yokozuna Hakuho's Behavior be Justified Ethically?

Hakuho simply wanted the referee to confirm on the video whether the *tachiai* initial charge had been properly completed or not. He did not refuse to obey the referee's judgment. In fact, roughly one minute later, he obeyed the referee's command to return to the ring. It is thus not difficult to imagine that the referee's compelling power had an effect on him.

If *Oh-sumo* is a modern sport, it would be clear that Hakuho's rights as a competitor have not been guaranteed. It goes without saying that the significance of a single victory is great for a competitor in professional sports. Based on the reactions of the media, NSK, and the public on SNS, however, it appears that contemporary Japan does not really want *Oh-sumo* to become a modern sport. If that is true, then we must conclude that Hakuho did not actually have the right to appeal the referee's decision.

Hakuho was criticized based on traditional values holding that it is undignified for a *yokozuna* grand champion to question a referee's judgment. How would the attacks on Hakuho that day appear to the eyes of a foreigner? Japan also has the Japan Sports Arbitration Agency, which is a division of the International Sports Arbitration Agency. Yet, due to the distaste for litigation in Japanese culture, the mass media treats athletes who simply seek arbitration as if they have done something wrong. Hakuho fully understood this aspect of Japanese culture.

The following day, when Hakuho was called before the judging department and given a severe warning, he immediately apologized. Was this the brave, wise, dignified, and virtuous thing to do? Hakuho apologized because he is accustomed to Japanese culture, understood it, and accepted national opinion.

Conclusion

The internationalization of sumo is already underway. The interest in sumo will probably increase further during the 2020 Tokyo Olympics.

Should *Oh-sumo,* Japan's national sport, really seek to become a modern sport? If the nation desires for it to continue in its current form as a hybrid sport, then it is fine as it is. If the idea is to internationalize sumo as a modern sport and an Olympic sport, however, it will be necessary to revise the rules, reflect the universal values of accountability and fairness, and protect the human rights of competitors.

Japanese *Oh-sumo* is at a crossroads. We argue that at the very least, even a hybrid sport must give consideration to the rights of athletes and protect the basic human rights that are guaranteed even in the Japanese Constitution. The state of Japanese *Oh-sumo,* which is caught between traditional sports and modern sports as illustrated so clearly by the Hakuho case, offers insight relevant to the status and future direction of traditional sports in many countries throughout the world.

References

1. Krieger D. In Sumo's Push for the Olympics, a Turn Away From Tradition. October 18th, 2010, New York Times.
2. Mainichi Newspaper, Hakuho 'Will sumo enter the Olympics?' Appeal to enter competition, October 4, 2016.
3. Svinth J. JAPAN: SUMO. In: Green TA and Svinth J (eds) Martial Arts of the World: An Encyclopedia of History and Innovation. 2016, pp.175-182.
4. Thompson L. Sport in Japan in the Early 21st Century: An Interpretation. In: Sport across Asia: Politics, Cultures, and Identities.London: Routledge, 2012, pp.102–111.
5. Guttmann A, Thompson L. Sumo, Ball Games, and Feats of Strength. In: Guttmann A and Thompson L Japanese Sport.Honolulu: University of Hawai'i Press, 2001, pp. 13-41.
6. Asahi Newspaper Digital. 2017a. November 22th, at 18:48. Defeated Hakuho, one minute spent appealing outside the ring, dissatisfied with *tachiai.*

7. Asahi Newspaper Digital. 2017b. November 22th, at 20:31, 'For someone who is a role model…' Match referees concerned by Hakuho's attitude.
8. NHK News Web, November 22th, 2017, at 20:32, Kyushu basho tournament 11th day, Hakuho's first defeat, claims 'false start.'
9. Asahi Newspaper Digital. 2017c. November 22th, at 23:34, So stubborn; the shameless behavior of a grand champion.
10. Asahi Newspaper Digital.2017d. November 23rd, at 21:28. The day after folly: Complete victory in 2 seconds, Hakuho: 'It wasn't drawn out.'
11. Asahi Newspaper Digital. 2017e. November 23rd, at 19:22, Hakuho apologizes: 'I am very sorry'; NSK Judging Department issues severe warning.
12. Mainichi Newspaper, November 23rd, 2017, at 20:59, Hakuho's objection provokes severe warning; Sumo Association Judging Department.
13. Simmonds NE. Introduction. In: Campbell D and Thomas P (eds), Fundamental Legal Conceptions as Applied in Judicial Reasoning by Wesley Newcomb Hohfeld,Burlington; Ashgate Publishing Company, 2001, xii-xiii.
14. Waldron J. Rights, In: A Companion to Contemporary Political Philosophy, Goodin R, Pettit P, Pogge T eds. 2nd ed., Oxford: Wiley-Blackwell, 2012, pp.746-747.
15. International Olympic Committee. Olympic Charter effective as of 9/5/2017, Chapter 6, Measures and Sanctions, Disciplinary Procedures and Dispute Resolution, 2017, pp. 88-92.
16. FIFA's Human Rights Policy – May 2017
17. Collins H. The Philosophy of Umpiring and the Introduction of Decision-Aid Technology. Journal of the Philosophy of Sport 2010; 37(2): p.136
18. Rawls, J.Two Concepts of Rules. Philosophical Review 1955; 64 (1): pp.24-29.
19. Simon RL. Competition, Selfishness, and the Quest for Excellence, In: Fair Play: Sports, Values, and Society, Boulder: Westview Press, 1991, p.21
20. Guttmann A. From Ritual to Record. In: From Ritual to Record:The Nature of Modern Sports, New York: Columbia University Press, 1978, pp. 15-55.
21. Thompson L. The modernization of sports and sumo. Review of Contemporary Sociology 1987; 24: 77-82. (in Japanese)
22. NHK (Nippon Hoso Kyokai). Official Sumo Rules. In: NHK ed. The Dictionary of Sports.Tokyo: NHK Publishing, Inc.; 1958: pp.77-88.,
23. Guttmann A. Cultural Imperialism, In: Games and Empires: Modern Sports and Cultural Imperialism, New York: Columbia University Press, 1994a, p.163.
24. Rummelt P. Sport im Kolonalismus-Kolonialismus im Sport (Cologne: Pahl-Rugenstein), 1986.
25. Eichberg H.Olympic Sport – Neocolonization and Alternatives. Int. Rev. for Soc. of Sport 1984; 19 (1): 97-105.
26. Guttmann A. Traditional Sport, In: Games and Empires: Modern Sports and Cultural Imperialism. New York: Columbia University Press, 1994b, p.188.

27. Sougawa T. Preface, In: Sougawa T ed., Traditional Sports in 21st Century, Tokyo: Taishu-kan, 2015, pp. i-v.
28. Rich M. Women Barred From Sumo Ring, Even to Save a Man's Life. April 5th, 2018, New York Times.
29. Tarrant J. Japan Sumo Chief Apologizes After Female Medics Asked to Leave Ring. April 5th, 2018, at 4:55, New York Times.
30. Asahi Newspaper Digital. 2018. April 6th, at 14:50. How vexing: A speech under the ring because of woman, Mayor Takarazuka City, Hyogo Prefecture.

Chapter 11
Rebirthing Bioethics: Going Global

Abstract This final chapter and the epilogue to follow revisit the idea of global bioethics. I begin by describing my experiences both as a member of the UNESCO International Bioethics Committee and in the field of international healthcare, challenging the current practices and academic discussion surrounding global bioethics, including that of global health. I cite the formative article by Van Rensselaer Potter as a way to question further the current state of the discussion concerning both the academic field and practice of global bioethics, arguing that the terms and the use therein all originate in the West. I explain my dissatisfaction with the discussion among researchers about the ongoing debate concerning universalism versus relativism. I then propose to discard the overarching term, "global bioethics," suggesting instead the use of a more challenging term that will fit future discussion. However, discarding the current terminology alone is insufficient. Therefore, I discuss what Japan could contribute to the establishment of new terminology. I conclude by reiterating the purpose of this book, calling for others worldwide to write their own versions, in order to facilitate a true international dialogue in bioethics.

This is the final chapter. Thus far, I have written about Japan as an example. However, when we shift to a global perspective on bioethics, what do we discover from this analysis? I hope to clarify this and the intent of this book once again at the end.

© The Author(s) 2020
A. Akabayashi, *Bioethics Across the Globe*,
https://doi.org/10.1007/978-981-15-3572-7_11

11.1   The UNESCO International Bioethics Committee

As a member of the International Bioethics Committee (IBC), I have had many opportunities to observe discussion at UNESCO meetings. Almost without exception, these situations tend to leave me feeling that advocacy for universal human rights was being presented superficially. UNESCO's the Universal Declaration on Bioethics and Human Rights (2005) states the following:

> Recognizing that ethical issues raised by the rapid advances in science and their technological applications should be examined with due respect to the dignity of the human person and universal respect for, and observance of, human rights and fundamental freedoms. (Author emphasis)

Meanwhile, the preamble to the UNESCO Constitution (1945) that serves as the premise of the Declaration states the following:

> That the great and terrible war which has now ended was a war made possible by the denial of the democratic principles of the dignity, equality and mutual respect of men, and by the propagation, in their place, through ignorance and prejudice, of the doctrine of the inequality of men and races….

> That the wide diffusion of culture, and the education of humanity for justice and liberty and peace are indispensable to the dignity of man and constitute a sacred duty which all the nations must fulfil in a spirit of mutual assistance and concern… (Author emphasis)

The use of Western expressions of human rights and justice at UNESCO conferences may not necessarily reflect the thinking of persons from African and Asian countries (at the very least, myself). While I value the activities of UNESCO, I do question the suitability of the methods and terminology used to achieve their goals.

I also question the future validity of this document. The Universal Declaration on Bioethics and Human Rights was created by a sub-organization of the UN, an organization created by countries representing the victors of World War II. To what extent is this declaration useful in the global development of bioethics? Is it truly capable of serving as a guideline?

The Preface of the Handbook of Global Bioethics (2014) states the following [1]:

> Preface

> A landmark in the early stage of global bioethics was the Universal Declaration of Bioethics and Human Rights, adopted by all member states of UNESCO (United Nations Educational Scientific and Cultural Organization), in 2005. This political and legal document presents the first general framework of ethical principles for global bioethics that covers all cultures and countries. It has been used as the major reference document for this Handbook. (Author emphasis)

A reading of the Preface signals some disquiet. Namely, the assumption that the Universal Declaration on Bioethics and Human rights should serve without question as a framework for "global bioethics."

Let me describe an experience I had while attending a UNESCO IBC conference. Although I was an individual committee member, I was also representing Japan. I had to make a courtesy call to the Japanese embassy in Paris and had dinner with the ambassador and governmental officials, where the ambassador and I exchanged opinions on current bioethical topics. After the dinner, the Foreign Ministry officials said something to the effect of "Dr. Akabayashi, tomorrow's conference topics have nothing to do with Japan, so you do not need to say anything, and if you do, you do not need to temper your words." Most bioethics topics require a global perspective, and thus, even if the topic do not directly relate to Japan, input from a Japanese perspective is still required. I was disappointed by this official's stance (i.e., "I'm not interested, because it's not my problem"), but also came to question the mission of the UNESCO IBC conference.

11.2 International Health (Global Health)

I wrote a commentary on a paper published by a researcher and activist from Africa who condemned her country's traditional practice of female genital mutilation (FGM) (Fig. 11.1) [2]. In my commentary, I mentioned that "it was time for (the author's) governments to intervene, and to establish clear policies on this issue." However, I also noted in my conclusion that before we criticize others, we must take a careful look at our own cultural practices.

> I have to confess that I am still not sure of the significance of discussing this issue as a researcher living in a different culture. What is the rationale for discussing these issues from outside? I agree with what Lane and Rubinstein stated in their previous discussion that:
>
> the method for moving beyond the impasse between cultural relativism and moral universals requires the careful, honest, and respectful conduct of conversations within our own society and between and among groups with different cultural suppositions.
>
> Some of the strongest critics to FGM are from so-called developed civilised countries. But how do such critics confront problems within their countries, such as serious issues in reproductive health/rights and child abuse? Before judging others, we should spend some more time understanding the flaws in our own culture.

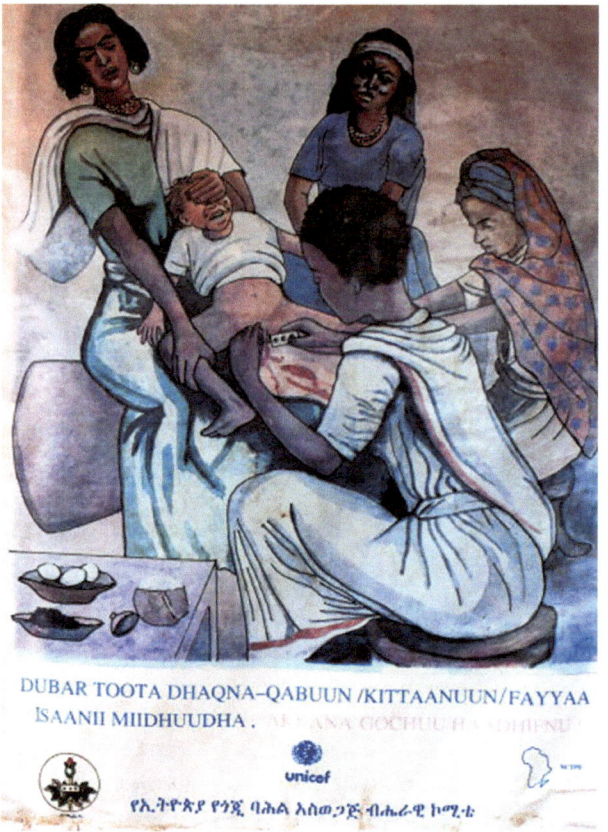

DUBAR TOOTA DHAQNA–QABUUN /KITTAANUUN/FAYYAA
ISAANII MIIDHUUDHA . ... ANA GOCHUU ... ...

unicef

የኢትዮጵያ የጎጂ ባሕል አስወጋጅ ብሔራዊ ኮሚቴ

Fig. 11.1 FGM

As another example of a global health issue, I often bring up the case described below in my international health bioethics class [3]. After introducing the background facts, I usually set up small discussion groups.

Case 18: Health Care Access & the Poor

Mrs. E is a 16-year-old female who is 39 weeks pregnant and lives in a small, poor village with a population of 200, roughly 400 km away from the nearest city (20 h by car). She lives with her family—a 17-year-old sister, her 22-year-old husband, and his parents. One morning, Mrs. E is found on the floor having a seizure and is later brought to the village's health center, which is run by Nurse P. Mrs. E's blood pressure measures 205/135 and protein is detected in her urine. She is diagnosed with severe eclampsia. According to Nurse P, the likelihood of mortality for Mrs. E and her baby is extremely high if nothing is done within the next 48 h. However, the family have neither a car nor enough money to pay for a ride to the city to seek proper treatment.

What Each Person Has to Say

Mrs. E: I just want my baby to be saved; I don't care if I die. Just save my baby.

Mr. E: Please save her and our baby. I'm willing to sell my kidney to save them.

Sister: I would even work in a brothel to make enough money to pay for their treatment.

Mr. E's parents: This has happened many times in the past in our village; it's sad but there is really nothing we can do about it.

Nurse P: I saw a similar case just last month. The girl ended up dying from severe sepsis during labor. It would be great to help Mrs. E and get her to the city for proper care.

Case Questions
1. Suppose you are a health care professional:
   What do you think Nurse P should do to help Mrs. E and her family?
2. Suppose you are a policymaker in your country:
   What policy would you propose to improve health care access in your country?

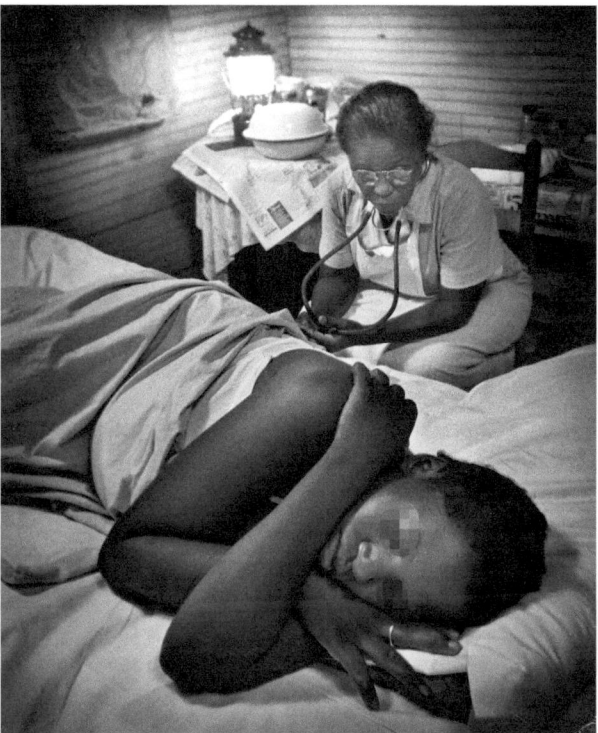

Photo 11.1 What to do, midwife Maude Callen?

Given the inherent complexities of the case, my class often end without any concrete conclusions. When handling more clearly defined topics such as abortion, surrogate childbirth, or euthanasia, the class discussions are fairly lively. However, when discussing this case, many students seem to be unable to find words to resolve the issues. Nonetheless, this experience pushes students challenging them to think from the relevant perspectives. The same is required in our discussion of global issues, particularly for topics related to health care disparities or, more directly, to distributive justice.

When I was working with Services for the Health in Asian and African Regions (SHARE), I participated in a project undertaken by the Japan International

Cooperation Agency (JICA) from Japan's Ministry of Foreign Affairs. We conducted field research on medical care policies in three African nations and offered detailed suggestions to the respective governments regarding Official Development Assistance (ODA). This project provided many opportunities to interview top governmental officials from each African country. However, one conclusion I came to, not only during the interviews with these officials, but also in my conversations with ambassadors and employees at Japanese embassies, was that regardless of how much ODA support was offered to LMICs by Japan, this funding would end up in the pockets of the supported country's government and intermediate organizations. Essentially, the funding was unlikely to reach the actual individuals in need.

When a professor from Thailand came to work in the Department of International Health Studies at the University of Tokyo, School of Medicine, I asked him: "Do you have any ideas about what would help support SHARE?" Without hesitation, this he responded, "Democratization!" I responded, "Would democracy be spread more readily by international health activity or ODA acts? What do you perceive as democracy? Do you see any limitations to democracy?" I received no reply from him.

Instead, this professor put a stop to the continuation of a US-Japan collaborative comparative research study on research ethics I (an assistant professor at the time) was conducting, refused to approve any opportunities for me to publish any papers, and our data from that study remains buried. Does democracy as defined by those from Thailand not allow for freedom of speech (which is allowed even by the Japanese Constitution)?

Having survived two World Wars and the Cold War, the mission of democracy no longer has any challenging force to keep it accountable. Meanwhile, the definition of democracy has become somewhat cloudy. As history reveals, many forms of democracy have existed, and include "democracy" as defined by the Thai people, Japanese, and many others. However, even the imprecise concept of "democracy" seems to be the current mainstream ideal. Given this reality, I believe that all we humans can do now is to see how far we can go with "revised" democracy. Democracy originated in ancient Greece and ancient Rome and boasts a long history of over 2000 years. If we intend to keep democracy alive by the year 3000, what can we do today to revise this ideology, and what does that even look like?

11.3   Van Rensselaer Potter, Inventor of Bioethics, his Acceptable Survival, and Anthropocentrism

In this chapter, I use the term "survival" frequently. I am drawing on of the concept of survival as coined by Van Rensselaer Potter, the Father of Bioethics. This also represents a return to the origin of bioethics. Potter [4–6] comments on five different forms of survival: "mere," "miserable," "idealistic," "irresponsible," and "acceptable." For example, "mere" survival implies food, shelter, and reproductive maintenance, but no progress beyond a more or less steady state. It implies no libraries, no written history, no cities, and no agriculture for urban support—essentially a "hunting and gathering" society. Potter's examples of this are the Inuit on the shores of the Arctic Ocean and the indigenous peoples of the Kalahari Desert in South Africa.

Potter proposes "acceptable survival" as the goal of global bioethics, defining it as "a long-term concept [1980] with a moral constraint: worldwide human dignity, human rights, human health, and a moral constraint on human fertility."

In response to the query "(a)cceptable survival for whom and acceptable to whom?" he notes that acceptable survival for all the world's people and acceptable to a universal sense of what is morally right and good and to what will realistically continue in the long term.

While Potter's argument considers the survival of the entire planet, it only incorporates perspectives from a portion of the West. If "mere survival" is excluded, then what will happen to the Inuit and Kalahari indigenous peoples? As Potter himself indicates in the same article, anthropologists would state that "to be a hunter and gatherer wasn't bad after all." Therefore, this perspective must require our respect. If we refer to them as "all the world's people," then are we claiming that if humanity does not seek acceptable survival for Inuit and Kalahari indigenous peoples as well, then will the human race will not survive to the year 3000? [7]

What then does Potter think about anthropocentrism, a concept borne from environmental ethics? I raise this because traditional anthropocentrism results in human overpopulation and progressive extinction of other species. On this matter, Potter states [8]:

> Neither were biocentrists, as Leopold might appear, nor anthropocentrists, as Otto might appear. Both are the ancestors and forerunners of *real* bioethics, although neither extrapolates to a consideration of organizational obligations in terms of what may now be called *real* bioethics: not biocentrism, not anthropocentrism, but a combination of both (p.181).

In other words, a confrontation between global bioethics and anthropocentrism is also needed. The warning about anthropocentrism is a necessary perspective for global survival, but if that call is issued, then can developed nations criticize the drilling of oil and deforestation needed for the economic development of LMICs? Can we criticize individuals in LMICs for having embraced anthropocentrism? As we continue in our failure to resolve issues such as global warming, should developed nations in the Western world actually be thinking about global survival?

In global bioethics discussions thus far, discussion on environmental ethics has been far from sufficient. Simply including some countries in Africa, Asia, and South America as authors in the Handbook of Global Bioethics does not guarantee that we will achieve a global perspective. After all, human rights, human dignity, justice, and biocentrism/anthropocentrism are all concepts derived from Western philosophy. If we never move even slightly away from this philosophical framework, then our thinking will never reach the neglected persons in LMICs. I am not advocating for the abandonment of the Western way of thinking, but I would welcome any new schemes yielding resolution of this issue, regardless of whether the developers are Western philosophers or not.

A new perspective in global bioethics is needed. Does global bioethics truly embody the conviction to do away with anthropocentrism? Can humankind actually construct a combination of both humanistic biocentrism and an enlightened nature-conscious anthropocentrism?

11.4   Universalism Versus Relativism

Universalism versus relativism is a long-standing debate in Western philosophy. However, in the 50 years in which global bioethics has evolved, the discussion has not moved forward in a meaningful way. In general, it appears that Western philosophers have agreed that the radical extremes (extreme universalism and extreme relativism) should be discarded, but that is all.

Moreover, several researchers have attempted to test the theory that revised universalism is akin to arguing for the lowest moral common denominator, adopting only that on which those from all cultures/religions/societies can agree [9–11]. Let us once again examine the contents of the Handbook of Global Bioethics [1].

> In the debate on globalization of ethics, Kymlicka (2007) has suggested that global ethics is a two-level phenomenon: at one level there is a self-standing international human rights discourse defining a minimum set of standards agreeable to all. At a second level, there is a multiplicity of different ethical traditions. These "local" traditions define what is ethically required beyond and above human rights. The same discussion can be used for global bioethics. On the one hand, there is a set of minimum standard on which traditions and culture agree; this is expressed in international human rights language and elaborated into specific bioethics principles. On the other hand, there are many efforts to articulate more specific bioethics standards in the context of specific religious and cultural traditions. (p.13)

If I were a Western philosopher, I likely would have followed this path. However, we must ask, to what extent have any of these academic challenges contributed or could contribute at all to the development of global bioethics in a practical manner?

Widdow and colleagues state the following [12]:

> However, to leave the paper here would leave unanswered, and more importantly unacknowledged, the questions of relativism and universalism that beset all areas of global ethics and indeed all areas of global study and practice. (p.110-111)

As a partial conclusion, they reject the through-going cultural relativist position and the exportation of Western individualistic values as if they were universal.

Nonetheless, this partial conclusion is nothing but a broken record, and their use of "beset" may even give the impression that Western authors are bound by an obsessive/compulsive curse. Thus I propose to Western philosophers that, "*only when thinking about the philosophical foundation of global bioethics, they must set aside terms involving the opposition between universalism and relativism,*" and work to construct a new framework of thinking. With regard to my question in the previous section ("Does global bioethics reject anthropocentrism?"), I believe that no matter how much we revise universalism and relativism, we cannot expect concepts worthy of serving as a philosophical foundation for global bioethics to emerge.

While it is fine to propose a revised universalism that discards anthropocentrism, we must ask: will such scholarly effort/work lead to any concrete behavioral guidelines? The "combination of both, a humanistic biocentrism and an enlightened nature-conscious anthropocentrism" mentioned in a previous section offers the exact same composition. Of course, I am not proposing that we set aside the framework of universalism versus relativism for all academic arguments in Western philosophy. However, they may come to realize that they have been using this as a premise; this may allow them to rethink this notion. I would like to think that this would lead to the envisioning of the next development.

In recent years, anthropology has come to be known as an academic field that assumes the stance of relativism, for which a major premise is the respect of different cultural values and plurality. In the past the field has assumed such a stance while failing to promote any norms, but that trend started to change in recent years, so that anthropology which strongly promotes norms is more common. In sociology as well, some researchers advocate strongly for norms. Meanwhile, the fields of ethics and moral philosophy—supposedly the classic fields of normative ethics— have come to acknowledge the importance of descriptive ethics and empirical ethics in the last 20 years, while in the field of bioethics, empirical ethics has gained wide acceptance. From a descriptive narrative approach, methodologies capable of arriving at normative conclusions are being developed. What does this all mean?

In the context of global bioethics, the framework defined by the classic opposition between universalism and relativism, and that between biocentrism and anthropocentrism, has reached a point where no further progress can be made, while the boundaries between universalism and relativism, and between biocentrism and anthropocentrism, are becoming unclear. Thus I advocate for the search for a new approach that fits the context of global bioethics. I will likely spend the rest of my life elucidating the specifics of such an approach, but I also know that its establishment is unlikely in my lifetime. Let us set aside the opposition between universalism and relativism in order to create a new framework, for if arguments pertaining to global bioethics continue in the same direction, then we will be forever stuck in a dead-end conversation.

11.5 Bioethics Across the Globe (BAG)

In a lecture as transcribed below, Campbell (2008) emphasized the following [13]:

> Perhaps, then, what we need to do is abandon the notion of "Asian Bioethics" as though this were some distinct and easily described entity. Instead we should discuss what might be important features of "Bioethics in Asia." This would suggest that we are dealing with the same discipline, but that the Asian context can add new dimensions, raise new questions or help to shift perspectives.

In the previous section, I suggested that we set aside the terms involving the opposition between universalism and relativism, so that we can construct a new framework of thinking. I alluded to this in the Introduction, but I also have a radical proposal. Namely, that we *discard* the term "global bioethics." This is not a proposal to set aside this term temporarily, with the potential for reuse later on, but rather to permanently reject it. As well, when using terms like "Asian Bioethics" and "African Bioethics," the adjectives "Asian" and "African" also signal relativism. Following Campbell in his lecture, I also believe we should no longer use the term "Asian Bioethics," as evidenced in my publication *"Biomedical Ethics in Asia: A Casebook for Multicultural Learners,"* where I have avoided using the term [3].

Therefore, I propose we use *"Bioethics Across the Globe (BAG)"* in place of "global bioethics." While I feel that this term embodies a more universal tone, I also wonder if some would interpret the adjective "global" (which already has a universal nuance) to mean that BAG is no longer trying to achieve universalism.

At the very least, *BAG* is preferable to the current term, "global bioethics." The time has come to *discard* the term "global bioethics," which will forever imply the opposition between universalism and relativism.

11.6 What Can Japan Contribute to BAG?

When Japan ended its period of national isolation in the early Meiji era, it temporarily played down all Japanese culture in order to usher in values and perspectives from the West. The German philosopher, Karl Löwith, who taught philosophy and German literature at Tohoku University in Japan from 1936 to 1941, accurately portrayed the situation in Japanese philosophy as follows:

> [Japanese people] are like those living in a two-story house, in which, on the first floor, they think and feel like Japanese, while in the second story all the various European scholarly texts, from Plato to Heidegger, are lined up, cover to cover. The European instructors question where the students are who go from one floor to the next. In actuality, they love themselves as they are. They have not yet eaten from the (proverbial Christian) tree of knowledge and have thus not yet lost their purity. They have yet to take the human out of themselves and have yet to taste the loss of human beings being critical of oneself. [14] (Der europaische Nihilismus, 1940).[1]

This text demonstrates that the Japanese attempted to study and understand modern perspectives of the West but had not accepted them in their entirety. The behavior exhibited here was a survival strategy for participating with Western powers. It was a strategy that Japan was forced to take in order to survive and thrive.

After Japan surrendered in World War II, it recovered economically under the protection of the USA. As Japan is a country with a predominantly single ethnicity,

---

[1] Sie leben wie in zwei Stockwerken: einem unteren, fundamentalen, in dem sie japanisch fühlen und denken, und einem oberen, in dem die europäischen Wissenschaften von Platon bis zu Heidegger aufgereiht stehen, und der europäische Lehrer fragt sich: wo ist die Treppe, auf der sie vom einen zum andern gehen? Im Grunde lieben sie sich so, wie sie sind, sie haben noch nicht vom (christlichen!) Baum der Erkenntnis gegessen und die Unschuld verloren, ein Verlust, der den Menschen aus sich herausstellt und ihn kritisch macht gegen sich selbst. (In German)

it did not have the added complication of racial conflict and was able to throw itself fully into economic growth. Nonetheless, there were also many values that were lost because of such growth, resulting, for example, in workaholism, working environments that are harsh enough to cause death from overworking, and the loss of identity involved in being a Japanese citizen.

As Japan entered the 1980s, bioethics was imported from the West in the name of "Global Standards," and Japanese scholars and practitioners themselves accepted these ways of thinking (e.g., self-determination) and societal systems (e.g., ethics review committees). This mirrored the experience from the Meiji era, just after national isolation had ended.

Having examined the ways in which Japan handled bioethics issues thus far, the reader may understand this strategy. Regardless of whether the issue involves brain-death and organ transplantation law, Japan's perspectives on the moral status of the human embryo, or something else altogether, the survival strategy of the Japanese people has not changed throughout history.

What then might Japan contribute to the development of BAG? It is one thing to be honest and vulnerable, without shame, so that others can truly understand. This honesty alone will promote the beginning of true dialogue. This book was not intended to be a critical assessment of Japan. I have portrayed Japan exactly as I perceive it. Japanese people reading this book should not reject this aspect of their identity (of course, each is free to declare an opposing opinion). We all live on the same planet, in different cultures and with different perspectives and values. These differences may not always be affirmed positively, but there is no need to justify one's own national stance; rather, we must begin by understanding these differences. This is how Japan "survives." This stance is the prerequisite for beginning a dialogue about BAG. I could have written solely about the beautiful, harmonious, parts of Japan, but there are already numerous publications that describe these. Every country has positive and negative aspects. My hope is that this book serves as a tool to promote that understanding. If Japan can first and foremost fulfill this role of promoting understanding by not denying the content of this book, it would be a very small first step, but will definitely contribute to BAG.

References

1. ten Have H, Gordijn B, editors. Handbook of global bioethics. Dordrecht: Springer Science + Business Media; 2014.
2. Akabayashi A. Female circumcision - a health issue or a human rights issue. Health Care Anal. 1998;6(1):55–8.
3. Akabayashi A, Kodama S, Slingsby BT. Biomedical ethics in Asia: a casebook for multicultural learners. Singapore: McGraw-Hill; 2010. p. 111–5.
4. Potter VR. Bioethics. Bridge to the future. Englewood Cliffs, N.J: Prentice-Hall; 1971.
5. Potter VR. Global bioethics: building on the Leopold legacy. Michigan: Michigan University Press; 1988.

6. Potter VR, Potter L. Global bioethics: converting sustainable development to global survival. Med Glob Surv. 1995;2:185–90.
7. Potter VR. Getting to the year 3000: can global bioethics overcome evolutions fatal flaw? Perspect Biol Med. 1990;34:89–97.
8. Potter VR. Real bioethics: biocentric or anthropocentric? Ethics Environ. 1996;1:177–83.
9. O'Neil O. Toward justice and virtue: a constructive account of practical reasoning. Cambridge: Cambridge University Press; 1996.
10. Barry B. Culture and equality. Cambridge: Polity Press; 2001.
11. Dickenson D. Risk and luck in medical ethics. Cambridge: Polity Press; 2003.
12. Widdow H, Dickenson D, Hellsten S. Global bioethics. New Review Bioeth. 2003;1(1):101–16.
13. Akabayashi A, Kodama S, Slingsby BT. Is Asian bioethics really the solution? Camb Q Healthc Ethics. 2008;17(3):270–2.
14. Löwith K. Der europaische Nihilismus. Betrachtungen zur geistigen Vorgeschichte des europaischen Krieges. Nachwort an den japanischen Leser. Stuttgart: J.B. Metzlersche Verlagsbuchhandlung; 1990. p. 111.

Epilogue

Finally, I will share a personal story.

I was born a mere twelve and a half years after Japan surrendered in World War II. No one can choose the place, time, or environment into which they are born. It was not a developed Western country or an LMIC in Africa, but as far as promoting the contents of this book, perhaps it was a "pretty good place." Or rather, perhaps it was precisely because I was born in Japan that I can make this argument.

Let me tell you about why I decided to become a physician. My father was a first lieutenant in the Japanese army. He graduated with top scores from military school and was part of an elite during World War II. However, after losing the war, he was forced to change his perspectives and values completely. He lost his job and suffered from maladaptation, choosing to follow the path of simply "living" in poverty. He chose to "survive." My grandfather, a principal at *Jinjo* Elementary school, imprinted Confucian morals on my mother, who then continued to instill in her own children those same teachings. Those Confucian morals were to "Be strict to yourself, be kind to others."

Having grown up in such an environment, I thought that "if I become a physician, then I can live comfortably in any era to come." Even if we were to go to war again, I could work as an army physician. Of course, I also had a strong desire to "serve the patients" (Confucian morals). However, equally important was my desire to survive.

In this book, I have been critical of the way in which that current global bioethics arguments had not moved beyond the framework of arguments within the Western philosophical paradigm. I even noted that no matter how far we move forward in this way, if we do not explore different understandings, we will never see any development in global bioethics. However, one limitation of this book is that it ends by criticizing the current state of the argument, but stops short of proposing a detailed alternative framework. *I want the reader to understand fully that proposing an alternative framework is not the intended purpose of this book.* Preparations to create a new framework have yet to begin in any country. I can safely say that no one in the world is currently able to propose a new framework.

The approach I took in a previous publication, *The Future of Bioethics: International Dialogues* was sound. It is not just a conversation. It is a true and deep dialogue that will not fall into relativism. However, Western and Eastern conversations have just begun. As such, when, and in what manner, will the true dialogue between West-East-North-South (for now, the delineations are merely geographical) on *BAG* begin?

I will say this again. Unfortunately, a *BAG* that accommodates the idea that humanity will still be thriving by the year 3000 has not as yet been born; in fact, it is currently in an embryonic or fetal state. Thus, in order to lead to the point of actual birth, we must begin our dialogue with the prize in sight, as this will serve as nutrition for the fetus and allow for growth and maturation until a healthy birth.

That dialogue begins with a mutual and deep understanding of others, which is the premise for true dialogue. This book is a tool that can be used to promote understanding, and which will serve as the basis for this dialogue. To return to the analogy of a fetus, this text is akin to the placenta. Is it not imperative that those of us still alive in 2020 look toward the birth of this fetus, working carefully and patiently to ensure the healthy and happy arrival of this child?

My final question is this: How should I live the remainder of my life? I cannot accept the vagueness of Kawabata. In Western terms, this is relativism. I ascribe to something much closer to Oe's universalism. However, the universalism spoken of by Oe is, in fact, vague. Many literary scholars and activists in Japan have proclaimed universalism, but they also failed in that they were only advocating the antithesis to Kawabata's vagueness, without fully understanding what exactly is universal.

No matter how much we proclaim vague universalism, this country will likely never change, because we have constructed and maintained this vague culture for over 2000 years. I think that Japan will continue to live an existence "suited to its own stature," and feel that it cannot accomplish anything other than this. Am I being pessimistic? Oe accepted his fate in this regard and committed to spending the remainder of his life in "healing." What shall I commit to do with my life?

My commitment from this point forward is to continue to serve as the placenta for the *BAG* embryo.

From the outside, Japan must seem like a truly mysterious and strange country. However, those who have taken the time to read to the end of this book will have hopefully deepened their understanding of Japan, its traditions, and culture. After all, my intention in writing this text was that readers would gain such an understanding. The present text is also a tool to help readers better understand a different culture with different values.

I also hope that this work will induce a chain reaction that leads to sequels from Western and non-Western countries; not just as country reports, but as true tools for dialogue. Once these resources and tools are all out on the table, our mutual understanding of each other will have deepened, at which point we will be able to, for the first time, begin a true North-East-South-West dialogue on bioethics.[1] To this end, I have entitled the book, *Bioethics Across the Globe: Rebirthing Bioethics.*

January 2020
Tokyo, Japan

---

[1] Why did I feel the need to write this book in English? The use of English as the common language has been criticized for various reasons. However, in 2020, in a time when the universal language of Esperanto has yet to be widely accepted, I have no other options. My previous text, *"Biomedical Ethics in Asia"* was written in simple English to facilitate its translation into multiple languages. My hope is for that text to be translated and used across various countries and languages. The English language is inevitably associated with values from the West. However, as there is currently no other way around this, and hardly any chance we will achieve a worldwide common language in my lifetime, for now, my message will remain in English, the language most likely to be understood by the most people around the world. If a universal language is developed in the future, I leave as my will that this book be translated into that language. …Will this be in the form of my living will or an advance directive?

Appendix

Akabayashi A, Hayashi Y.
Informed Consent Revised: A Global Perspective.
In: Akabayashi A, editor. The future of bioethics: international dialogues.
Oxford: Oxford University Press; 2014. p. 735–49.

Introduction

Informed consent is fundamentally grounded in the idea of respect for individual autonomy. However, health-care workers, patients, and their families sometimes encounter complicated situations regarding the interpretation and practice of informed consent. This is particularly true in cases of poor prognosis and the subsequent decisions regarding care. To illustrate the difficulty implementing informed consent as it has been traditionally interpreted, two cases, one from Japan and one from the USA, were presented by one of the authors in previous publications [2, 3]. We proposed that a family-facilitated approach to informed consent could be equally appropriate to the first-person approach that has been commonly regarded as the standard method of ensuring patient autonomy.

The family-facilitated model of informed consent was discussed earlier as an alternative to the first-person approach rather than as a replacement for it. This reconceptualization of informed consent, while adding some complexity, takes into account cultural norms and different ways of thinking about the self. Specifically, a patient with an independent self-construction is one who identifies her/himself as an autonomous individual with an independent set of values, and is thus highly consistent with the conventional paradigm of autonomy and the standard approach to informed consent. In contrast, we suggest that many patients have an interdependent view of themselves, thus perceiving themselves as more connected and responsive to others. Thus, the patient with an interdependent self-construal is likely to identify her/himself first as a component of the family unit, in which case the family-facilitated approach may be more appropriate.

An article published in the 2006 American Journal of Bioethics (AJOB) by one of the authors (in collaboration with others) addressing this issue provoked a variety of reactions. Many of the responses shared some common misinterpretations of our position. We believe there were two reasons for these misunderstandings. Firstly, the premises for the argument promoting the use of the family-facilitated approach were not adequately communicated. Secondly, we asserted that the family-facilitated approach is compatible with respect for autonomy but did not adequately explain the reasoning behind the assertion. In this paper, we provide a more detailed description of the proposed family-facilitated approach to informed consent and present arguments establishing that this approach is not inconsistent with patient autonomy.

In the following two sections, we will review the previously published cases and clarify the characteristics of a family-facilitated approach. To further explain the family's role in proxy decision-making and differentiate the family-facilitated approach from other models of informed consent, we present an additional case that took place in Japan in 2010. We then go on to develop new models of informed consent. We call these the "soft proxy" approach and the "hard proxy" approach. We argue that the family-facilitated approach is supported by a form of autonomy that differs from some conventional views. This argument invokes the idea of tacit consent by a patient, as the family-facilitated approach does not require explicitly expressed consent. We defend the family-facilitated approach by describing a form of autonomy that is consistent with the original intent as governance of the self. We argue that the family-facilitated approach excludes the possibility of paternalism by the physician and undue influence by the family. We reassess the relationship between the family-facilitated approach and these new models in light of relational autonomy. Finally, we review the philosophical basis of the family-facilitated approach, and suggest that it is applicable to other topics. We assert that the first-person approach and family-facilitated approach to informed consent can be useful in many cultural settings beyond the borders of Japan and the USA.

Case I: A Japanese Patient in the Late 1980s

Case I was previously published in the Journal of Medical Ethics (JME).[1] The patient, while still healthy, had indicated to her family that she did not want to know the diagnosis if she ever developed cancer. At the time of her diagnosis the family requested that the physician not disclose the news to the patient.

---

[1] The full description of Case 1 is as follows [3: 296].

A 62-year-old Japanese woman presented to a Tokyo hospital with a fever and severe back pain. Diagnostic work-up included serological tumor marker testing and abdominal computed tomography. This revealed advanced gall bladder cancer metastatic to the liver and back. Since her expected survival was less than three months and she was not a candidate for surgery or chemotherapy, a regimen of comfort measures and pain control was needed. The diagnosis was first revealed to her family members, namely her husband and her son, separately from the patient. The husband and son discussed it with the daughter, and together the family requested that the patient not be told. The family explained that while still healthy the patient had mentioned to them her wish not to be told if she developed cancer. This mention of her preference may have been stimulated by intermittent media coverage of the issue in Japan and seemed plausible. After initial treatment for pain and fever, the patient stabilized and was competent to participate in decision-making,

The Custom of First Notifying the Patient's Family of a Diagnosis With Poor Prognosis

In Japan, until the 1980s, bad news such as a diagnosis of advanced-stage cancer was delivered first to the family rather than to the patient, because the family was thought to be in the best position to understand the patient's ability to comprehend the diagnosis and make the best decisions to promote his or her welfare. Physicians and family members were also often concerned that the patient would be traumatized by the diagnosis. Acknowledging the importance of family support, physicians often sought the family's consent before disclosing a cancer diagnosis to patients, even when the patient was clearly competent [3]. This approach is not exclusive to Japan. In Italy, for example, bad news was also disclosed to the family, not to the patient [16].

"Something Close to Autonomy" and Its Implications

"Family consent, communication, and advance directives for cancer disclosure: a Japanese case and discussion," discusses the idea of "something close to autonomy," as described by the North American medical ethicist Edmund Pellegrino. Pellegrino argues that autonomy, or something close to autonomy, is a universal principle, not just a cultural artifact. However, he challenges the commonly held view of first-person consent as essential through his description of cases in which a patient is not apprised of his/ her actual diagnosis, and shows how this nonetheless might be consistent with a form of autonomy.

In many cultures clinicians encounter patients who are fully aware of the gravity of their condition but choose to play out the drama in their own way. This may include not discussing the full or obvious truth. This is *a form of autonomy*, if it is *implicitly and mutually* agreed on, between physician and patient. [13: 1735; italics added]

He continued thus:

> Autonomy is still a valid and universal principle because it is based on what it is to be human. The patient must decide how much autonomy he or she wishes to exercise, and this amount can vary from culture and culture. It seems probable that the democratic ideals that lie behind the contemporary North American concept of autonomy will spread and that *something close to it* will be the choice of many individuals in other countries as well. [13: 1735; italics added]

---

though she was a little withdrawn and dependent. The treating physician and family met with the patient and in the family's presence, the treating physician told her: "You don't have any cancer yet, but if we don't treat you, it will progress to a cancer." In response, the patient asked for no further details. An aggressive pain-control regimen was continued and though she was intermittently drowsy, she died four months later without apparent suffering from physical pain. The physician never explicitly discussed the diagnosis with her.

What Pellegrino is suggesting is that patient autonomy or something close to it is applicable across cultures, and moreover that this particular notion of autonomy should be respected because it is based on patient preference. However, he rejects the strict interpretation of autonomy used in North America as universal, instead recognizing diverse beliefs about the interpretation and implementation of autonomy in different cultures.[2]

In this context, the fundamental question becomes: How is autonomy understood in other countries?

Our analysis of Case 1, described above, concludes that the patient's autonomy was respected based on the following criteria: (I) the patient's prior declaration of her wishes; (2) the physician's vague explanation, which did not reveal the cancer diagnosis to the patient, respected the patient's wishes; and (3) the patient, who was competent and had the opportunity to question the physician but chose not to do so. However, when discussing the case in question we did not discuss specifically what form of autonomy was being respected and in what way.

Case 2: An American Patient in the 1990s

Case 2, previously presented in AJOB, involved a patient in the USA.[3] This is a case in which, at the time of the onset of cancer, the patient's family asked the physician not to disclose the diagnosis to the patient. The patient was alert, oriented, and competent; therefore, it is not clear why the family received information before the patient herself. Because the practice of informing the family first is not typical in the USA, we assume this is a case involving an ethnic minority.

The nursing staff were not willing to tell the patient her diagnosis without the attending physician's permission (and the patient had never asked, though she was upset about being in the hospital). The attending said that he had seen patients for over 30 years and that if the patient did not ask what was wrong, he would not tell them. He thought that patients "usually figure out what is wrong anyway and adjust

---

[2] Further arguments discussed in the JME article are as follows...[W]hen considering this issue in the international context, the term "autonomy" should be used carefully since it is not a concept with only one meaning. Pellegrino does not specify whether his notion of a North American concept of autonomy refers to the definition of autonomy or the degree of exercise of autonomy, or both. Surbone's remark that autonomy is often synonymous with isolation in Italy illustrates that the exercise of autonomy differs in Italy and North America, even though the definition may be very similar [3: 299].

[3] The full description of Case 2 is as follows [2: 9]. A 74-year-old woman was admitted for increased blood sugar and fever. A CT scan revealed multiple liver masses. A biopsy revealed a squamous cell carcinoma. The patient's family (a daughter and two sons) was told the diagnosis and insisted that the patient not be told. They were afraid that the knowledge would decrease her will to live and thus shorten her life. The patient was very close to her family as she spent most weekdays with her daughter and the weekends at home with her unmarried son. By all accounts, the patient was alert and oriented.

quite well." The attending said that he understood that it was "in fashion" to tell patients what is wrong with them, but that he disagreed with this silly trend.

The family became extremely upset when again approached about informing the patient and questioned the justification for the "hospital policy" that patients should be told their diagnosis.

Defining the Family-Facilitated Approach and Its Premises

In the paper referred to above the family-facilitated approach was proposed as an alternative to the traditional first-person approach to informed consent.

A family-facilitated approach to informed consent where family and patient function as a single unit differs from the more popular first-person approach. In this paper, we define a family-facilitated approach as a process of informed consent in which a patient's family communicates with the attending physician and medical staff and often makes treatment-related decisions. This differs from acting as a proxy in that the patient does not officially appoint his or her family. [2: 11]

The validity of a family-facilitated approach is based on two premises: (I) a patient-family fiduciary relationship, and (2) a patient who identifies her/himself more as a component of the family unit than as an independent individual. This kind of fiduciary relationship is one in which decisions regarding the patient as made by the family are assumed to be in their own best interests.

Self-construction and the Binary Approach

The binary model, in which both the first-person approach and family-facilitated approach can be used, takes into account divergent concepts of self-construction-independent and interdependent. A person with an independent view identifies her/ himself as an autonomous individual with an individual set of values and a unique perspective.[4] In the health-care setting, a patient with an independent view would naturally prefer a first-person approach to informed consent—free to make his or her own decisions based on careful consideration of the fully disclosed risks and benefits of each treatment option. In contrast, an individual with an interdependent self-construction will tend to identify her/himself as part of a larger unit consisting of family, friends, and others.[5] Accordingly,

---

[4] The independent view was originally described as follows: This view of the self derives from a belief in the wholeness and uniqueness of each person's configuration of internal attributes ...The essential aspect of this view involves a conception of the self as an autonomous, independent person [12: 226].

[5] The interdependent view was originally described as follows: This view of the self and the relationship between the self and others features the person not as separate from the social context but as more connected and less differentiated from others [12: 226].

such patients may be more comfortable with a collaborative process in which disclosure of risks and decision-making regarding treatments involve important members of that unit. In Case 2, it is apparent that the patient had a more interdependent view of self and therefore she would have preferred a family-facilitated approach.[6]

The Relationship to Autonomy in Case 2

Our discussion of Case 2 in our previous paper concluded that the family-facilitated approach was not inconsistent with respect for patient autonomy.[7] However, as with Case 1, we did not describe the philosophical basis for this conclusion. Both cases show situations in which the family plays an important role in decision-making, even though the patient did not explicitly designate any particular family member as his or her legal representative. However, in our earlier work we did not explain why the family is entitled to perform such a role in such situations, or how the family-facilitated approach differs from use of a legal proxy. We will clarify these points below.

Case 3: Mr. K, a Japanese Patient in 2010

Thus far, we have summarized some arguments for recognizing the validity of the family-facilitated approach in two previously published case studies. In this section, we present a third case to further elucidate the characteristics of the family-facilitated approach. Although fictitious, this example is based on an actual case from 2010.

---

[6] Explanation in the article was as follows: In this case, we need to ask whether or not the patient was willing to entrust her decision-making to her family. This depends largely, however, on her relationship with her family and her self-construal. The patient was "alert and oriented" (competent), "very close with her family," and she never asked her attending physician about her diagnosis. Judging by these facts, if she had not been willing to entrust her decisions to her family, she more than likely would have asked her family or attending physician directly about her disease. Deducing from the fact of her competence combined with her silence on the matter intimates that she indeed held an interdependent view [2: 12].

[7] The analysis in the article was as follows: Moreover, a family-facilitated approach does not necessarily contradict with the general ethical principle of respect for autonomy in the USA. In fact, a family-facilitated approach to informed consent may be respecting a patient's individual choice. That is, if a patient who holds an interdependent view has a propensity to prefer a family-facilitated approach, providing this approach to informed consent may indeed be respecting patient autonomy [2: 13].

Case Presentation: Mr. K

Mr. K was a 61-year-old corporate executive who presented with cervical adenopathy in 2010. Stage 4 metastatic squamous cell cancer was diagnosed by biopsy. The prognosis was unclear, but assumed to be very poor because the cancer was of unknown primary origin. Mr. K was informed of the diagnosis by the physician and presented with a choice of chemotherapy, radiation therapy, or surgery. His wife was present during the consultation. Mr. K had been married for more than 30 years and had an excellent relationship with his wife. They had no children. Mr. K's siblings included an older sister and younger brother; however, both were married and he was not close to either of them.

Mr. K told his physician and wife, "I am shocked. I cannot make this decision by myself." This reaction and apparent loss of decision-making capacity made his wife uneasy, and she turned to the physician for advice. However, the physician would not recommend a specific treatment, replying that the patient and family must make the decision. The wife consulted with a family member who was a medical practitioner, and gathered information from the Internet. After careful consideration, she arrived at the following conclusion: Begin treatment with chemotherapy and then, if necessary, pursue radiation therapy or surgery. After consulting with the physician to confirm the appropriateness of her decision, Mrs. K told her husband in the presence of the physician, "I spoke with the doctor and decided to start with chemotherapy. I hope that's OK," to which Mr. K nodded, and, without inquiring about the details of the chemotherapy or the pros and cons of the treatment alternatives, he was subsequently admitted to the hospital for treatment.

A New Type of Informed Consent in Contemporary Japan

Mr. K's case illustrates a new way of thinking about informed decision-making in contemporary Japan. Our analysis is based upon three central points. Firstly, Mr. K says that he is shocked by the diagnosis. Nonetheless, he exhibits no sign of difficulty making other types of decisions such as giving work-related instructions to those at his company. In that sense, he is competent enough to participate in decision-making regarding his treatment. However, he explains that he regresses when emotionally taxed, and therefore cannot make these difficult medical decisions on his own.[8] Secondly, Mr. K and his wife have a long-standing relationship and get along well. Thus, we may assume that a fiduciary relationship exists between him and his wife as a family unit. Because he explains that he is unable to make the decision alone and wants to rely on others, it is reasonable to assume that he holds an interdependent view of himself. Finally, Mr. K nods in agreement with his wife's

---

[8] The human tendency to withdraw in difficult situations or weakened conditions is an important psychological defense mechanism necessary for survival.

explanation of her decision regarding his treatment. Mr. K was given the opportunity to disagree with his wife's decision and refuse hospitalization or treatment.[9]

Cases like the one described above whereby a family member takes leadership in the decision-making on behalf of a competent patient still occur. The question remains: What type of approach is appropriate for a patient who is emotionally withdrawn? Mr. K has shown that he has an interdependent view of himself. The two conditions described above required for a family-facilitated approach are in place. Therefore, we consider it appropriate that his wife, in consultation with the physician, makes the therapeutic decisions on his behalf.

The interdependent view does not compel strict self-determination for a patient who is emotionally withdrawn. In contrast, a strictly first-person approach requires that, unless the patient is clearly incompetent and requires legal representation, the patient must take responsibility for decision-making based on comprehension and consideration of fully disclosed information.

The Use of Proxy Decision-Making in Case 3

Mr. K did not officially appoint his wife as a proxy. His wife saw that her withdrawn husband was not able to make important decisions about his own treatment. She therefore used her own judgment based on her understanding of her husband's best interests to choose a treatment strategy after consultation with his physician and finding additional information on her own. She informed her husband by telling him, "I talked with the doctor and decided to start with chemotherapy. I hope that's OK." The husband just nodded without apparent interest in the process by which his wife reached the decision and without requesting any details about the treatment options.

Although Mr. K approved his wife's decision, he did not officially appoint her as a proxy. We wish to use the term "soft proxy" approach for the way in which Mr. K's wife is entitled to make decisions in such a case. The soft proxy approach can be defined in the following way: despite no official request by the patient, the family and physician decide on a treatment strategy independently of the patient, and this decision is ultimately confirmed by the patient. The soft proxy approach is congruent with the Japanese idea of *omakase* (patient leaving the decision to others, especially family members or physicians). In contrast, a conventional autonomous proxy

---

[9] In Japan, enforcement of a new Personal Information Protection Act in 2003 enabled patients to obtain information from their medical records. It is reasonable to assume that Japanese physicians have directly informed patients of their diagnoses in almost all cases since that time. However, results of surveys have revealed that the decision to fully disclose the actual prognosis to a patient is still left to the best judgment of the health-care worker. According to the survey, physicians tended to disclose the patient's prognosis to his/her family pessimistically, but reveal the same prognosis to the patient rather optimistically. See Akabayashi et al. [1].

is a third party, such as a lawyer, who is explicitly and officially appointed by the patient. This "hard proxy" approach is based on the assumption of an independent perception of the self. The process of integral family involvement in informed consent and decision-making may be easier to understand through these concepts of soft proxy approach and hard proxy approach.

In What Sense Is the Family-Facilitated Approach Consistent or Inconsistent With Patient Autonomy?

So far we have described what we call the family-facilitated approach in the context of Cases I and 2. We believe that a family-facilitated approach is appropriate in cases such as these where patients have an interdependent self-construal and the two above-mentioned requirements are met. We have also suggested that taking a family-facilitated approach in such cases is not inconsistent with patient autonomy. However, some might believe that the approach is in conflict with a strict interpretation of patient autonomy, which would require direct discussion with patients, an assessment of their competence, full disclosure of information, and respect for patient self-determination based on adequate comprehension of the benefits, burdens, and risks of all reasonable alternatives. Some may argue that patient autonomy in this strict sense was not adequately respected in the two cases described above. In this section we defend the family-facilitated approach by considering arguments asserting that a family-facilitated approach is consistent with respect for patient autonomy.

Is the Family-Facilitated Approach Compatible With the Conventional View of Autonomy?

To see why we believe a family-facilitated approach is compatible with patient autonomy, let us revisit our conclusions from our analysis of Case 2. In Case 2, the argument was: if a patient who holds an interdependent view has a propensity to prefer a family-facilitated approach, providing this approach to informed consent may indeed be respecting patient autonomy [2: 13]. The line of reasoning behind this argument begins with our assumption that patient autonomy is being respected when a patient's preferences are fulfilled. We then argue that a patient with an interdependent view of himself is highly likely to be more comfortable with a family-facilitated approach, thus taking a family-facilitated approach is consistent with the patient's preference. Therefore, we conclude, a family-facilitated approach is consistent with the patient's autonomy in that it is in accord with the patient's preferences.

We believe that it is safe to assume that patients in cases such as these two examples can be assumed to have interdependent view of themselves and prefer that the family make the decisions. The patient in Case 2 did not consult the physician despite the opportunity to do so, nor did she oppose her exclusion from the decision-making process. The patient in Case I was also given the opportunity to consult the physician and did not refuse the involvement of the family in the decision-making process. Thus, in both cases, the patient tacitly consented to the family making the medical decisions. In effect the patient ultimately authorized this process. Thus, the family-facilitated approach is consistent with the preference of patients who have an interdependent view of themselves.[10]

When consent is tacit, some doubt may remain as to whether patient autonomy can be said to be truly respected. Autonomy as conventionally interpreted requires explicit, rather than tacit, consent. We shall return to this question below.

A Comparison of Four Models for Informed Consent With Regard to Patient Autonomy

To address the potential concerns stated in the previous paragraph, as well as to see how our conclusions could be applied in broader contexts, it will prove helpful to make a brief summary of the landscape as described thus far. Let us now review the four models that have been presented (Fig. A.1).

The first-person approach is a decision-making process with expressed consent and is in accordance with the conventional view of patient autonomy. It is consistent with an independent self-construal. When applying the hard proxy approach, the patient does not actively make the medical decisions, but officially and explicitly appoints a legal representative of his/her own choosing. The legal proxy may be a third party outside the family, such as an attorney. In this case, the patient is also likely to have an independent construction of him/herself. His/her relationship to the legal representative is a fiduciary one.

By contrast, the soft proxy approach used in Case 3 is consistent with an interdependent construal of the self and satisfies the two premises needed for a family-facilitated approach. In addition, a close relative can act as the representative without being explicitly designated as such by the patient. Here, decisions made by the

---

[10] By comparison, in Case 3 it may seem that the wife consulted with the physician and made the medical decision without Mr. K's permission. However, despite being able to confer with his wife or the physician, he did not do so. In effect, he did not use his veto power. Moreover, he provided explicit rather than tacit authorization by nodding. In Case 3, the conventional view of autonomy supports either (I) that the patient should make decisions himself or (2) that the legal representative has to be designated by the patient with explicit authorization (in which case the scenario would fit with the hard proxy approach model).

	Self-construal	Premise 1 fiduciary relationship	Premise 2 Identification as part of a unit	Conception of autonomy	Degree of ideal vs. local practice respected
First-person approach	Independent	Not necessary	Not necessary	Completely conventional	Ideal practice (respect for conventional autonomy)
Hard proxy approach	Independent	Required	Not necessary	Completely conventional	
Soft proxy approach	Interdependent	Strongly required	Required	Somewhat conventional (omakase)	
Family-facilitated approach	Interdependent	Strongly required	Strongly required	"Something close to autonomy"	Local cultural practice (respect for family-oriented decisions)

Fig. A.1 Four models for informed consent

family are ultimately confirmed by the patient. There is some expression of consent by the patient. In the family-facilitated approach, the decisions are made by the family without explicit designation of proxy by the patient, and the patient undergoes the chosen treatment, giving only tacit consent. The family-facilitated approach is consistent with the notion of an interdependent construal of self and requires the two premises of patient-family fiduciary relationship and patient primary identification as part of a family unit to be satisfied.

The differences among the four models can be summarized as follows: (1) whether the patient or another party makes decisions with regard to specific treatments (first-person approach vs. all others), (2) whether the patient provides tacit or expressed consent (family-facilitated approach vs. all others), and (3) the degree of voluntariness when deciding the level of proxy (soft proxy approach vs. hard proxy approach).

When we consider the issue of autonomy within these four approaches, the first-person approach and hard proxy approach are consistent with the idea of respecting patient autonomy as traditionally understood. The same cannot be said for the soft proxy approach and family-facilitated approach. However, because the patient expresses their consent in the soft proxy approach, it does not lack due respect for patient autonomy as understood traditionally. The family-facilitated approach requires only tacit consent, and therefore may appear not to respect the patient's autonomy. Thus, the family-facilitated approach is not acceptable within the scope of this conventional view of autonomy.

This analysis might strengthen questions about whether or not patient autonomy is respected in the family-facilitated approach. Our aim in the following section will be to address the way in which the family-facilitated approach can be compatible with patient autonomy.

What Sort of Autonomy Is Compatible With the Family-Facilitated Approach?

In what sense can we claim that the family-facilitated approach is still compatible with patient autonomy? In answering this question, we wish to emphasize that in the family-facilitated approach, physician paternalism and any undue influence from the family are excluded as points of contention. In the family-facilitated approach, the patient's desire for family decision-making, authorized by tacit consent, is respected, and the possibility that the physician will make decisions against the will of the patient is removed. It is important to remember that the rejection of physician paternalism is the original issue in contemporary bioethics regarding respect for patient autonomy. Thus, the family-facilitated approach addresses arguments criticizing potential paternalism.

It is true that strong family involvement in medical decision-making may appear oppressive and in conflict with patient autonomy. However, in the family-facilitated approach, it is the patients who want their family to make the medical decisions, as they see themselves first as part of a family unit, and family decision-making is preferred by the patients themselves. Therefore, family involvement is not an undue restriction to patient autonomy.

Based on these considerations we conclude that the family-facilitated approach is compatible with the motive behind the conventional view of autonomy, although the family-facilitated approach is not compatible with autonomy in the strictest conventional sense of the word. However, we maintain that it is consistent with some particular sort of autonomy, which fits well with the Japanese clinical settings and any other settings where the family's role in treatment choice is considered more significant, including certain Italian, Chinese, and some American subcultures. This is congruent with Pellegrino's expression "something close to autonomy," "a form of autonomy." Pellegrino, who first stated that a form of autonomy was used in the case of a patient in Italy, did not offer a clear definition [13: 1735]. We claim that the crux of what Pellegrino calls "something close to autonomy," "a form of autonomy" might best be understood as the minimization of physician paternalism and respect for patient preference.

Relational Autonomy and the Family-Facilitated Approach

To further clarify the concept of autonomy in the family-facilitated approach, it is useful to compare it to the concept of relational autonomy. According to Catriona Mackenzie and Natalie Stoljar "[t]he term 'relational autonomy'… does not refer to a single unified conception of autonomy but is rather an umbrella term, designating a range of related perspectives" [11: 4]. However, these perspectives share "the conviction that persons are socially embedded and that agents' identities are formed within the context of social relationships and shaped by a complex of intersecting social determinants, such as race, class, gender, and ethnicity" (ibid).

Stated simply, the exploration of relational autonomy is characterized as an attempt to facilitate the reconsideration of contemporary philosophical accounts of autonomy based on this view of identity formation. As Susan Sherwin stated: "Under a relational view, autonomy is best understood to be a capacity or skill that is developed (and constrained) by social circumstances. It is exercised within relationships and social structures that jointly help to shape the individual while also affecting others' responses to her efforts at autonomy [15: 36]."

Reexamination of the concept of autonomy from a relational perspective has a range of implications for bioethical problems. We will focus on the arguments of Anita Ho, which address the connection between respect for patient autonomy and family influence from a relational standpoint. By focusing attention on the patient's vulnerability and relational identity, she suggests that we need to reevaluate the role of family and respect for patient autonomy in situations of medical decision-making.[11]

Patient's Consent, Family's Role, and Relational Autonomy

According to Ho [9], when a patient's relational identity is understood, family involvement is not necessarily in conflict with respect for patient autonomy. In fact, in certain cases it may actually enhance it. In the health-care setting, patients can feel isolated and in such circumstances they may (I) prefer preservation of identity through family connections to self-determination, and (2) prioritize intimate relationships and the welfare of loved ones over their individual interests. Respecting the autonomy of the patient may be considered equivalent to respecting the patient's needs and wishes, which are influenced by the family.[12]

The fundamental issues explored in the study of relational autonomy by feminist bioethicists correspond with the core philosophical issues upon which the family-facilitated approach is based. First, the family-facilitated approach is founded on a construction of self similar to that found in writings about relational autonomy. Both consider

---

[11] For other attempts to discuss the role of family in decision-making in connection with the idea of relational autonomy, see Lee [10] and Turoldo [17].

[12] Ho argues as follows. "In this context of parenthood, dependency or family involvement may preserve rather than violate autonomous agency—it can help to maintain a range of identifications that can promote patients' own sense of integrity and worth. Because it is reasonable to assume that intimates generally care deeply about the patient's interest and well-being, such that their choices would probably match the patient's overall goals, family involvement can be compatible with or even enhance the patient's autonomy" [9: 131]. Further, she argues: "For those whose family is at the centre of their existence, consideration of their advice, needs and mutual interests is part of their autonomous agency" [9: 132].

carefully the importance of human relationships. In addition, the goal of the family-facilitated approach to reevaluate the conventional definition of autonomy and pursue alternatives shares a great deal with the feminist analysis of relational autonomy. Finally, Ho acknowledges that the family has an important role in medical decision-making, and asserts that family intervention does not necessarily infringe respect for patient autonomy. The family-facilitated approach also recognizes this assertion.

Oppression and the Feminist Interpretation of Autonomy

Nonetheless, we disagree with Ho's interpretation of respect for patient autonomy when she describes what kind of patient consent is needed. We have argued that within the family-facilitated approach tacit consent is the minimum requirement of respect for patient autonomy. Ho, on the other hand, still considers the patient's explicit expression of wishes a strict condition of respect for autonomy. In her commentary on the AJOB article in which the Case 2 was originally published, she states:

I argue that, in cases where patients defer decision-making to family members, unless there is clear evidence of neglect or abuse, caregivers should follow the patients' expressed wishes. [9: 26; italics in the original]

Additionally, in another paper, Ho further contends that "professionals' default position should be to trust the patient's own final expressed wishes" [9: 133]. While Ho believes it is important that some patients prefer preservation of identity through family connections to self-determination, she nonetheless insists upon expressed consent.

What underlies Ho's insistence on the expressed intent of the patient, we believe, is concern about the potential for oppression that is shared by other feminist writers. Indeed, Ho states that "One concern that may arise here is that patriarchal and other oppressive relationships often disguise manipulation, exploitation, control and abuse as love and familial bond" [9: 133]. Requiring the patient's expressed consent may therefore be the last safeguard against manipulation and exploitation.

Expressed Consent or Tacit Consent: Do They Truly Differ?

The concern that oppression gives rise to manipulation and exploitation is well founded. We suspect, however, that expressed consent may not be sufficient in itself for removing the potential risk of manipulation and exploitation of a patient with a relational identity. If in fact it is impossible to save the patient from exploitation and undue influence of the family based on their tacit consent to medical procedures, then it should prove to be impossible to do so with the patient's expressed consent as well.

To understand this point, let us compare Case 3 (the soft proxy approach), in which the patient expresses his consent, with Case 2 (the family-facilitated approach), in which there is only tacit consent. Although the patient in Case 3 expressed his consent to treatment by nodding, this alone does not ensure that there is less risk of exploitation or undue pressure from the family than in Case 2. For this to occur, more is needed such as the empowerment of the individual exposed to oppression, as suggested by Susan Sherwin (15: 34–39). However, in situations requiring medical decisions, such decisions cannot wait for the patient's empowerment. While there are complicated social and institutional issues that make it very difficult to realize the empowerment of individuals in need, medical problems need concrete solutions appropriate to the circumstances and are subject to time and resource constraints.

Hence, if we take seriously the reality of such severe clinical situations like those presented in the cases under discussion, we conclude that there is no substantial difference between obtaining the patient's expressed consent and acknowledging the patient's tacit consent when considering the patient's vulnerability to exploitation or undue pressure from the family.

Informed Consent Revised: A Global Perspective

The arguments above suggest that the unique conditions of a clinical setting as well as the relationships of a patient placed therein have to be taken into consideration in establishing what we call "a form of autonomy" (i.e. a concept of autonomy for which obtaining the patient's expressed consent is not an essential criterion). Of course, even in this instance, the patient's wishes are made known by his tacit consent. It is for this reason that we consider the family-facilitated approach as "a form of autonomy."

In developing this family-facilitated approach model, we sought to balance respect for patient autonomy and the cultural importance of the family in decision-making. Restrictions of time and resources often influence the process of medical decision-making, and there are situations in which the ideal practice of informed consent is not possible or may be inappropriate. In addition, problems directly related to life and death demand immediate solutions. Thus, we acknowledge the importance of taking this reality into account in order to arrive at practical and ethical solutions.

Using the example of informed consent, we acknowledge the significance of the family's role in decision-making in local cultures. On the one hand, we attempt to pay due respect to local cultural values to the extent that they are compatible with the concept of autonomy that underlies the ideal practice of informed consent. On the other hand, we expand the interpretation of autonomy while shedding light on the interdependent construal of self that underlies the value of family decision-making in local cultural practices. Through attempting to reconcile apparently conflicting abstract ideals and local realities without giving either of them absolute status, we have developed the family-facilitated approach as a solution.

Such complexity cannot be reduced to a simple algorithm. Nonetheless, this method can provide a starting place for practical solutions that avoid the pitfalls of parochial ethnocentrism and arrogant universalism. This may provide a methodology for developing better solutions in a progressively globalized world.

The methodology and philosophy described here may also be applicable to other problems in clinical/medical ethics, given that hard questions in medical ethics often involve conflicts among universal and local values. For example, the process can be used to examine how to deal with different definitions and criteria of death or to reveal the merits and problems of living donor organ transplantation.

Using informed consent as an example, we developed four models of informed consent, which have here been discussed in the context of Japan and the US. We have argued that the family-facilitated approach is not a replacement for the first-person approach. Patients should be able to choose the approach best suited to them, and this may change for the same individual over time.

It is possible that these models could become choices for patients outside Japan and the US. Further, we hope our flexible methodology will be applied in different regions and cultures. This in turn may engender other informed consent models that are sensitive to the diversities of clinical realities?.[13] As Pellegrino says, "this question of balancing autonomy is today a necessary part of any transcultural dialogue in medical ethics. The ethics of medicine offers a fruitful point for beginning a larger cultural dialogue between and among the world's major cultures" [14: 18]. We believe that our models and our way of thinking described here mark an important milestone in the reassessment of informed consent on a global scale.

Fan and Tao [7] also suggest that theoretical explanations for the family-determination model of informed consent may be provided by appealing to the concept of tacit consent.

The difficulty is to provide an adequate moral-theoretical account of the role of families and physicians in decision-making [in China and Hong Kong].... For contemporary bioethics, the theoretically least challenging approach is to account for the role of the physician and the family *by appealing to a background authorizing tacit consent.* [7: 141–2; italics added]

Nevertheless, rather than choosing to provide explanations that appeal to tacit consent, Fan and Tao defend the family-determination model by highlighting the importance of the family's role in decision-making within the cultural context of Confucianism. Their unique views of the concept of autonomy may underlie this approach. According to Fan [4], the Western conception of autonomy as a "self-determination-oriented principle" and East Asian conception of autonomy as a

---

[13] For example, in a series of papers, Ruiping Fan and his colleagues employ the Confucian perspective to defend a family-determination model of informed consent that is familiar within a Chinese clinical context [5, 6, 7, 18]. This family-determination model is similar to the family-facilitated approach in that it recognizes situations in which it is more appropriate for the family, rather than the patient, to make decisions, regardless of the patient's decision-making capacities.

"family-determination-oriented principle" are disparate concepts; "there is a different principle of autonomy implicit in the cultural and ethical traditions of East Asian countries which is incommensurable with Western principle of autonomy [4: 313]." Rather than expanding the concept of autonomy as we do, Fan attempts to explain the propriety of family decision-making through the importance of the family's role from a Confucian perspective.

Thus, the conclusions of Fan et al. are similar to our conclusions, but the approaches taken to arrive at these conclusions differ. Whereas Fan et al. enthusiastically argue the importance of the family's role in decision-making based on Confucian concepts, we argue that decision-making by the family can be interpreted as respect for patient's autonomy using the interdependent construal of self. Notably, the concept of autonomy in this case refers to 'a form of autonomy, an alternative concept of autonomy that stands alongside the conventional construction of autonomy that underlies first-person approach.

We have not aggressively demonstrated or defended the importance of the family's role, but this should not be interpreted as a denial of the family's importance. Rather, we choose to refrain from making direct moral judgments regarding the importance of the family's role, and argue that the family-facilitated approach gains support from the principle of autonomy as long as the concept of autonomy can be understood in the way explained in this paper.

References

1. Akabayashi A, et al. Truth telling in the case of a pessimistic diagnosis in Japan. Lancet. 1999a;354(9186):1263.
2. Akabayashi A, Slingsby BT. Informed consent revisited: Japan and the U.S. Am J Bioeth 2006;6(1):9–14.
3. Akabayashi A, Fetters MD, Elwyn TS. Family consent, communication, and advance directives for cancer disclosure: a Japanese case and discussion. J Med Ethics. 1999b;25:296–301.
4. Fan R. Self-determination vs. family-determination: two incommensurable principles of autonomy. Bioethics. 1997;II(3–4):319–22.
5. Fan R. Informed consent and truth telling: the Chinese Confucian moral perspective. HEC Forum. 2000;12(1):87–95.
6. Fan R, Li B. Truth telling in medicine: the Confucian view. J Med Philos. 2004;29(2):179–93.
7. Fan R, Tao J. Consent to medical treatment: the complex interplay of patients, families, and physicians. J Med Philos. 2004;29(2):139–48.
8. Ho A. Family and informed consent in multicultural setting. Am J Bioeth. 2006;6(1):26–8.
9. Ho A. Relational autonomy or undue pressure? Family's role in medical decision-making. Scand J Caring Sci. 2008;22:128–35.

10. Lee SC. On relational autonomy: from feminist critique to Confucian model for clinical practice. In: Lee SC, editor. The family, and medical decision-making, and biotechnology: critical reflections on Asian moral perspectives. Dordrecht: Springer; 2007. p. 83–93.

11. Mackenzie C, Stoljar N. Introduction: autonomy refigured. In: Mackenzie C, Stoljar N, editors. Relational autonomy: feminists perspectives on autonomy, agency, and the social self. New York: Oxford University Press; 2000. p. 3–31.

12. Markus HR, Kottayam S. Culture and the self: implications for cognition, emotion, and motivation. Psychol Rev. 1991;98:224–53.

13. Pellegrino ED. Is truth telling to the patient a cultural artifact? JAMA. 1992a;268(13):1734–5.

14. Pellegrino ED. Prologue: intersections of western biomedical ethics and world culture: problematic and possibility. In: Pellegrino ED, Mozzarella P, Corse P, editors. Transcultural dimensions in medical ethics. Frederick, MD: University Publishing Group; 1992b. p. 13–9.

15. Sherwin S. A relational approach to autonomy in health care. In: Sherwin S, editor. The politics of women's health: exploring agency and autonomy. Philadelphia, PA, Temple University Press; 1998. p. 19–47.

16. Sorbonne A. Truth telling to the patient: letter from Italy. JAMA. 1992;268(13):1661–2.

17. Turoldo F. Relational autonomy and multiculturalism. Camb Q Healthc Ethics. 2010;19:542–9.

18. Wang M, Lo P-C, Fan R. Medical decision and the family: An examination of controversies. J Med Philos. 2010;35(2):493–8.